Ship Design and Performance for Masters and Mates

Ship Design and Performance for Masters and Mates

Dr C.B. Barrass

ELSEVIER

BUTTERWORTH
HEINEMANN

AMSTERDAM BOSTON HEIDELBERG LONDON NEW YORK OXFORD
PARIS SAN DIEGO SAN FRANCISCO SINGAPORE SYDNEY TOKYO

Elsevier Butterworth-Heinemann
Linacre House, Jordan Hill, Oxford OX2 8DP
200 Wheeler Road, Burlington, MA 01803

First published 2004

British Library Cataloguing in Publication Data
A catalogue record for this book is available from the British Library

Library of Congress Cataloging in Publication Data
A catalog record for this book is available from the Library of Congress

ISBN 0 7506 6000 7

For information on all Elsevier Butterworth-Heinemann publications
visit our website at http://books.elsevier.com

Typeset by Charon Tec Pvt. Ltd, Chennai, India
Printed and bound in Great Britain

Contents

To my wife Hilary and our family

Acknowledgements

I gladly acknowledge with grateful thanks, the help, comments and encouragement afforded to me by the following personnel of the Maritime Industry:

Steve Taylor, UK Manager, Voith Schneider Propulsion Ltd.

Jörg Schauland, Becker Marine Systems, Hamburg.

Tim Knaggs, Editor, Royal Institute of Naval Architects, London.

Graham Patience, Managing Director, Stone Manganese Marine Limited, Birkenhead.

Lyn Bodger, Technical Manager, Stone Manganese Marine Ltd., Birkenhead.

John Carlton, Lloyds Surveyor, Lloyds Registry in London.

Paul Turner, Retired Fleet Manager (Engine & Deck side), P&O Ship Management.

Captain Neil McQuaid, Chief Executive, Marcon Associates Ltd., Southport.

Captain Tom Strom, Director, Cunard Line Ltd/Seabourn, Cruise Line Miami.

Introduction

The main aim is to give an introduction and awareness to those interested in *Ship Design and Ship Performance*. It is written to underpin and support the more erudite books published on Naval Architecture and Marine Engineering by Elsevier Ltd.

It will also bring together the works of Masters, Mates, Marine Engineers and Naval Architects engaged in day-to-day operation of ships at sea and in port.

Part 1   This part illustrates how a ship is designed from limited information supplied from the shipowners to the shipbuilders. It shows how, after having obtained the Main Dimensions for a new ship, the Marine Engineers select the right powered engine to give the speed requested by the shipowner in the Memorandum of Agreement.

Chapter 1 deals with determining the Main Dimensions. Chapter 2 looks into how group weights are estimated. Chapters 3 and 4 analyse capacities and hydrostatics for new vessels.

Personnel engaged in the Maritime Industry can sometimes be uncertain on which *resistance*, which *speed* or which *power* is being referred to in meetings. Chapters 5–8 will assist in removing any such uncertainty. Chapter 9 shows preliminary methods for designing a propeller and a rudder for a new ship.

Part 2   Chapters 10 and 11 give particulars relating to modern Merchant ships. After a ship has been designed and built, she must then be tested to verify that the ship has met her design criteria. She must attain the shipowner's prerequisites of being seaworthy and commercially viable. Chapters 12–16 cover the various ship trials carried out by the shipbuilder on a newly completed ship.

Over the last three decades, ships have greatly increased in size (e.g. Supertankers). They have also increased in service speed (e.g. Container ships). Groundings and collisions have become more common. Frequently this has been due to ship squat and Interaction effects. One only has to recall the incidents of 'Herald of Free Enterprise', and the 'Sea Empress'.

Chapters 17–19 explain these problems. Suggestions are given for reducing the effects of excessive squat and interaction.

Occasionally errors in design do result. Chapters 20 and 21 discuss in detail, how shortfalls can be put right, with either a replacement or with a retrofit.

Chapter 22 discusses the improvements in propeller performance.

This book tabulates general particulars of 39 ships designed, built and delivered in this Millennium. It also covers many ship types designed and built over the last 20 years. Discussed in detail are new inventions and suggestions for enhanced ship performance in the next decade.

Finally, if you are a student, good luck in your studies. If you are either sea-going or shore-based personnel, best wishes for continued success in your job. I hope this book will be of interest and assistance to you. Thank you.

Dr. C.B. Barrass

Part 1
Ship Design

Chapter 1

Preliminary estimates for new ships: Main Dimensions

It has been said that the problem for a Naval Architect is to design a ship that will carry a certain deadweight at a reasonable rate of stowage in a seaworthy vessel at a predetermined speed on a given radius of action as cheaply as possible all in conjunction with a General Arrangement suited to the ship's trade.

The Naval Architect must therefore keep in mind all of the following:

- Main Dimensions
- Hull form
- Displacement
- Freeboard
- Depth
- Capacities
- Trim and stability
- Economic considerations

- Longitudinal and transverse strength
- Structural scantlings
- Resistance and powering
- Machinery
- Endurance
- Wood and Outfit
- Lightweight and deadweight
- Material costs

In determining the Main Dimensions for a new ship, guidance can be taken from a similar ship for which basic details are known. This is known as a 'basic vessel' and must be similar in type, size, speed and power to the new vessel. It is constantly referred to as the new design is being developed.

When a shipowner makes an initial enquiry, he usually gives the shipbuilder four items of information:

- Type of vessel
- Deadweight of the new ship
- Required service speed
- Route on which the new vessel will operate

The intended route for a new vessel is very important for the designer to know. For example there may be a maximum length to consider. If the new vessel is to operate through the Panama Canal her maximum length must be 289.56 m. For the St. Lawrence Seaway the restriction for length is 225.5 m.

Beam restriction for the Panama Canal is 32.26 m and 23.8 m for the St. Lawrence Seaway. Draft restriction for the Panama is 12.04 m up to the tropical fresh water mark. For the St. Lawrence Seaway the draft must be no more than 8.0 m. For the Suez Canal, there are limitations of ship breadth linked with Ship Draft.

Finally there is the Air Draft to consider. This is the vertical distance from the waterline to the highest point on the ship. It indicates the ability of a ship to pass under a bridge spanning a seaway that forms part of the intended route. For the Panama Canal, this is to be no greater than 57.91 m. For the St. Lawrence Seaway the maximum Air Draft is to be 35.5 m.

The first estimate that the Naval Architect makes is to estimate the lightweight of the new ship. Starting with some definitions:

1. *Lightweight*: This is the weight of the ship itself when completely *empty*, with boilers topped up to working level. It is made up of steel weight, wood and outfit weight and machinery weight.
2. *Deadweight*: This is the weight that a ship *carries*. It can be made up of oil fuel, fresh water, stores, lubricating oil, water ballast, crew and effects, cargo and passengers.
3. *Displacement*: This is the weight of the volume of water that the ship displaces. Displacement is lightweight (lwt) + deadweight (dwt). The lightweight will not change much during the life of a ship and so is reasonably constant. The deadweight however will vary, depending on how much the ship is loaded.

Deadweight coefficient C_D: This coefficient links the deadweight with the displacement:

$$C_D = \frac{\text{deadweight}}{\text{displacement}} = \frac{\text{dwt}}{W}$$

C_D will depend on the ship type being considered. Table 1.1 shows typical values for Merchant ships when fully loaded up to their Summer Loaded Waterline (SLWL) (Draft Mld). The abbreviation Mld is short for moulded.

Table 1.1   Typical dwt coefficients for several Merchant ships

Ship type	C_D@SLWL	Ship type	C_D@SLWL
Oil Tanker	0.800–0.860	Container ship	0.600
Ore Carrier	0.820	Passenger Liners	0.35–0.40
General Cargo ship	0.700	RO-RO vessel	0.300
LNG or LPG ships	0.620	Cross-channel	0.200

As a good first approximation, for General Cargo ships and Oil Tankers, it can be stated that at the SLWL, the C_B approximately equals the C_D where:

$$C_B = \frac{\text{volume of displacement}}{L \times B \times H}$$

where:

L = Length between perpendiculars (LBP),
B = Breadth Mld,
H = Draft Mld.

Worked example 1.1

For a new design, a shipowner has specified a dwt of 9000 tonnes. Information from a database of previously built similar ships suggests C_D to be 0.715.
    Estimate the fully loaded displacement (W) and the lwt for this new ship.

$$C_D = dwt/W \quad \text{So} \quad W = dwt/C_D$$

$$W = 9000/0.715 = \quad 12\,587 \text{ tonnes}$$

$$-dwt \text{ (as specified)} = \quad -9000 \text{ tonnes}$$

$$lwt = \quad \underline{3587 \text{ tonnes}}$$

The dwt coefficient is not used for Passenger vessels. This is because dead-weight is not so important a criterion. Furthermore, Passenger vessels are usually specialist 'one-off ships' so selection of a basic ship is much more difficult. For Passenger vessels, floor area in square metres is used as a means for making comparisons.

Estimations of the length for a new design

1. Ship *length* is controlled normally by the space available at the quayside.
2. Ship *breadth* is controlled by stability or canal width.
3. Ship *depth* is controlled by a combination of draft and freeboard.
4. Ship *draft* is controlled by the depth of water at the Ports where the ship will be visiting. Exceptions to this are the ULCCs and the Supertankers. They off-load their cargo at single point moorings located at the approaches to Ports.

Method 1: Cube root format

From information on ships already built and in service, the Naval Architect can decide upon the relationships of L/B and B/H for the new ship.
    Knowing these values he can have a good first attempt at the Main Dimensions for the new vessel. He can use the following formula:

$$L = \left(\frac{dwt \times (L/B)^2 \times (B/H)}{p \times C_B \times C_D} \right)^{1/3} \text{ m}$$

where:

L = LBP in metres for the new ship,
B = Breadth Mld in metres,
p = salt water density of 1.025 tonnes/m³,
C_B and C_D are as previously denoted.

Worked example 1.2

From a database, information for a selected basic ship is as follows:

$$C_D = 0.715, \quad C_B = 0.723, \quad L/B = 7.2, \quad B/H = 2.17$$

For the new design the required dwt is 6700 tonnes. Estimate the L, B, H, lwt and W for the new ship.

$$C_D = dwt/W \quad \text{So} \quad W = dwt/C_D$$

$$W = 6700/0.715 = \quad 9371 \text{ tonnes}$$

$$-dwt \text{ (as given)} = -6700 \text{ tonnes}$$

$$\text{lwt} = \quad 2671 \text{ tonnes}$$

$$L = \left(\frac{dwt \times (L/B)^2 \times (B/H)}{p \times C_B \times C_D} \right)^{1/3} \text{ m}$$

$$= \left(\frac{6700 \times 7.2 \times 7.2 \times 2.17}{1.025 \times 0.723 \times 0.715} \right)^{1/3}$$

$$= \underline{112.46 \text{ m}}$$

$$L/B = 7.2 \quad \text{So} \quad B = L/7.2 = 112.46/7.2 = 15.62 \text{ m}$$

$$B/H = 2.17 \quad \text{So} \quad H = B/2.17 = 15.62/2.17 = 7.20 \text{ m} = \text{SLWL}$$

Check!!

$$W = L \times B \times H \times C_B \times p$$

$$W = 112.46 \times 15.62 \times 7.2 \times 0.723 \times 1.025$$

$$W = 9373 \text{ tonnes (very close to previous answer of 9371 tonnes)}$$

These values can be slightly refined and modified to give:

L = 112.5 m, B = 15.60 m, H = 7.20 m, C_D = 0.716, C_B = 0.723,
fully loaded displacement (W) = 9364 tonnes, lwt = 2672 tonnes.

In the last decade, LBPs have decreased in value whilst Breadth Mld values have increased. The reasons for this are threefold.

Because of oil spillage following groundings, new Oil Tankers have double skins fitted. These are formed by fitting side tanks P&S, where it is hoped they will reduce loss of oil after side impact damage. In essence, a form of damage limitation.

Alongside this has been the development of Container ships with the demand for more deck containers. Some of these vessels are large enough to have 24 containers stowed across their Upper Deck.

In order to reduce vibration and strength problems together with decreases in first cost, Oil Tanker designers have tended to reduce the LBP. To achieve a similar dwt, they have increased the Breadth Mld. L/B values have gradually reduced from 6.25 to 5.50 to 5.00.

One such vessel is the 'Esso Japan' with 350 m LBP and a Breadth Mld of 70 m, and a massive dwt of 406 000 tonnes. Truly an Ultra Large Crude Carrier (ULCC). Another example is the 'Stena Viking' delivered in April 2001. She has a dwt of 266 000 tonnes, an LBP of 320 m and a Breadth Mld of 70 m. This makes her L/B a value as low as 4.57.

Method 2: The geosim procedure

This is a method used when a new order is geometrically similar to a basic ship. The method is as follows.

Worked example 1.3

A 100 000 tonnes dwt Very Large Crude Carrier (VLCC) is 250 m LBP, 43 m Breadth Mld and 13.75 m Draft Mld. Her C_B is 0.810 and her C_D is 0.815.

A new similar design is being considered, but with a dwt of 110 000 tonnes. Estimate the new principal dimensions, W and the corresponding lwt.

For geosims $(L_2/L_1)^3 = W_2/W_1$

Thus $L_2/L_1 = (W_2/W_1)^{1/3} = (111\,000/100\,000)^{1/3}$

$L_2/L_1 = 1.0323 = \text{say } K$

New LBP = old LBP × K = 250 × 1.0323 = 258.08 m

New Breadth Mld = old Breadth Mld × K = 43 × 1.0323 = 44.389 m

New draft = old draft × K = 13.74 × 1.0323 = 14.194 m.

Check!!

$$W = L \times B \times H \times C_B \times p$$

$$W = 258.08 \times 44.389 \times 14.194 \times 0.810 \times 1.025$$

$$W = 135\,003 \text{ tonnes}$$

$C_D = \text{dwt}/W = 110\,000/135\,003 = 0.8148$ say 0.815, same as the basic ship.

$$\text{lwt} = W - \text{dwt} = 135\,003 - 110\,000 = 25\,003 \text{ tonnes}$$

Dimensions could be refined to L = 258 m, B = 44.4 m, H = 14.2 m.

The main drawback with this method is that it only serves as a first approximation, because it is unlikely in practice that:

$$L_2/L_1 = B_2/B_1 = H_2/H_1 = K$$

Finally note that for both vessels $C_B = 0.810$ and $C_D = 0.815$.

Method 3: Graphical intersection procedure

From a study of a large number of Merchant ships, it has been shown that in modern ship design practice, the parameters L and B can be linked as follows:

$B = (L/10) + (5\ \text{to}\ 7.5)\ \text{m}$	General Cargo ships
$B = (L/10) + (7.5\ \text{to}\ 10)\ \text{m}$	Container vessels
$B = (L/5) - 12.5\ \text{m}$	Supertankers (C.B. Barrass 1975)
$L/B = 6.00\text{--}6.25$	Supertankers (1975–1990)
$L/B = 5.00\text{--}5.75$	Supertankers (1990–2004)

C_B can also be linked with service speed (V) and the LBP (L) in that:

$$C_B = 1 - m\,(V/L^{0.5}) \qquad \text{Evolution of Alexander's formula.}$$

The slope 'm' varies with each ship type, as shown in Figure 1.1. However, only parts of the shown straight sloping lines are of use to the Naval Architect. This is because each ship type will have, in practice, a typical design service speed.

For example, an Oil Tanker will have a service speed of say 15–15.75 kt, but generally not more than 16 kt. A General Cargo ship will have a service speed in the order of 14–16 kt but normally not greater than 16 kt. A Container ship will be typically 20–25 kt service speed, but not less than 16 kt. Further examples are shown in Table 1.2.

Table 1.2   Typical $V/L^{0.5}$ values for several Merchant ships

Ship type	Typical fully loaded C_B value	Typical service speed (kt)	LBP circa (m)	$V/L^{0.5}$ values
VLCCs	0.825	15.50	259.61	0.962
Oil Tankers	0.800	15.50	228.23	1.026
General Cargo ships	0.700	14.75	132.38	1.282
Passenger Liners	0.625	22.00	222.77	1.474
Container ships	0.575	22.00	188.36	1.063

Figure 1.1 shows C_B plotted against $V/L^{0.5}$. It shows Alexander's straight line relationships for several ship types, with the global formula suggested by the author in 1992. This global formula can replace the five lines of previously plotted data. The equation for the global formula is:

$$C_B = 1.20 - 0.39\,(V/L^{0.5}) \qquad \text{C.B. Barrass (1992)}$$

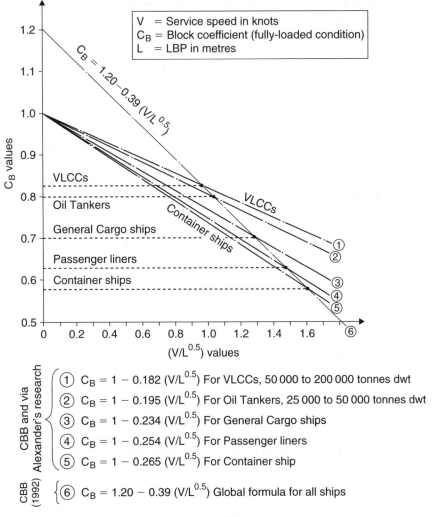

Fig. 1.1   Graphs of $C_B \propto V/L^{0.5}$ for several ship types.

The content within the figure:

V = Service speed in knots
C_B = Block coefficient (fully-loaded condition)
L = LBP in metres

C_B values (y-axis), $(V/L^{0.5})$ values (x-axis)

$C_B = 1.20 - 0.39 (V/L^{0.5})$

VLCCs
Oil Tankers
General Cargo ships
Passenger liners
Container ships

VLCCs
Container ships

CBB and via Alexander's research
CBB (1992)

① $C_B = 1 - 0.182 (V/L^{0.5})$ For VLCCs, 50 000 to 200 000 tonnes dwt
② $C_B = 1 - 0.195 (V/L^{0.5})$ For Oil Tankers, 25 000 to 50 000 tonnes dwt
③ $C_B = 1 - 0.234 (V/L^{0.5})$ For General Cargo ships
④ $C_B = 1 - 0.254 (V/L^{0.5})$ For Passenger liners
⑤ $C_B = 1 - 0.265 (V/L^{0.5})$ For Container ship
⑥ $C_B = 1.20 - 0.39 (V/L^{0.5})$ Global formula for all ships

Worked example 1.4

A ship has an LBP of 124 m with a service speed of 14.25 kt.

(a) Estimate C_B at her fully loaded draft.
(b) If a new design of similar length but with a speed of 18 kt, what would be
    her C_B value?

(a) $C_B = 1.20 - 0.39 (V/L^{0.5})$
    $C_B = 1.20 - 0.39 (14.25/124^{0.5})$
    $C_B = 0.700$

(b) $C_B = 1.20 - 0.39 (18.00/124^{0.5})$
    $C_B = 0.570$

The first ship is likely to be a General Cargo ship. It is quite likely that the second ship is a RO-RO vessel.

Generally, it can be assumed that the higher the designed service speed, the smaller will be the corresponding C_B value. As we increase the design service speed, the hull contours will change from being full-form (Oil Tankers) to medium-form (General Cargo ships) to fine-form (Container vessels).

Worked example 1.5

The Main Dimensions for a new vessel are being considered. She is to be 14000 tonnes dwt with a service speed of 15 kt, to operate on a maximum summer draft of 8.5 m.

Estimate LBP, Breath Mld, C_B and W if from basic ship information, the C_D is to be 0.700 and B is to be (L/10) + 6.85 m.

$$W = dwt/C_D = 14\,000/0.700 \quad \text{So} \quad W = 20\,000 \text{ tonnes}$$

$$W = L \times B \times H \times C_B \times p \quad \text{So} \quad C_B = W/(L \times B \times H \times p)$$

$$C_B = 20\,000/\{L \times (L/10 + 6.85) \times 8.5 \times 1.025\}$$

$$= 2295.6/\{L \times (L/10 + 6.85)\} \tag{1}$$

$$= 1.20 - 0.39\,(V/L^{0.5}) \quad \text{as per global formula}$$

$$= 1.20 - 0.39\,(15/L^{0.5})$$

$$= 1.20 - 5.85/L^{0.5} \tag{2}$$

Now equation (1) = equation (2)

Solve graphically by substituting in values for L.

Let L = say 142 m, 148 m and 154 m, then C_B values relation to LBP values are given in Table 1.3.

Table 1.3 C_B values relating to LBP values

Length L (m)	C_B	
	Equation (1)	Equation (2)
142	0.768	0.709
148	0.716	0.719
154	0.667	0.729

Figure 1.2 shows the two sets of C_B values plotted against the LBPs. When the two graphs intersect it can be seen that C_B was 0.718 and L was 147.8 m.

$$L = 147.8\,m$$

$$\text{Breadth Mld} = (L/10) + 6.85 = 14.78 + 6.85 = 21.63\,m$$

$$H = 8.5\,m, \text{ as prescribed in question.}$$

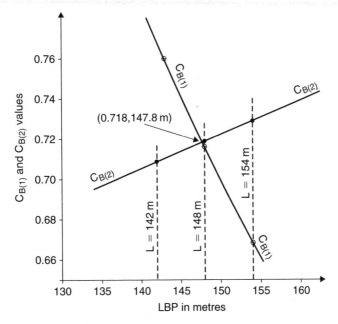

Fig. 1.2 C_B values against LBP values for Worked example 1.4.

$$W = L \times B \times H \times C_B \times p$$

$$= 147.8 \times 21.63 \times 8.5 \times 0.718 \times 1.025$$

$$= 19\,999 \text{ tonnes} \qquad \text{say } 20\,000 \text{ tonnes, as previously calculated.}$$

After modifying and slightly refining:

$$L = 148 \text{ m}, \quad B = 21.60\,\text{m}, \quad H = 8.5\,\text{m}, \quad C_B = 0.718, \quad C_D = 0.700,$$

$$W = 20\,000 \text{ tonnes}, \quad \text{lwt} = W - \text{dwt} = 20\,000 - 14\,000 = 6000 \text{ tonnes.}$$

Selection of LBP values for graphs

Collection of data from various sources suggest the approximate values given in Table 1.4. These values were plotted and are shown in Figure 1.3.

Table 1.4   General Cargo ships:
approximate LBP against dwt

Approx LBP (m)	Deadweight (tonnes)
97.6	4000
112.8	6000
125.0	8000
134.2	10000
143.5	12000
151.0	14000

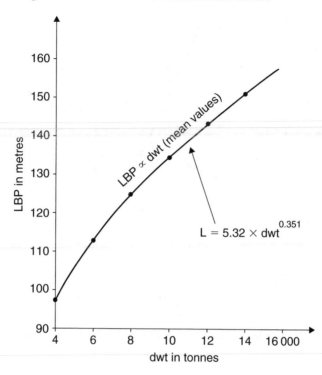

Fig. 1.3   LBP ∝ dwt for General Cargo ships.

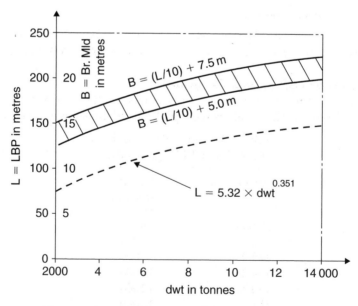

Fig. 1.4   (L and B) ∝ dwt for General Cargo ships.

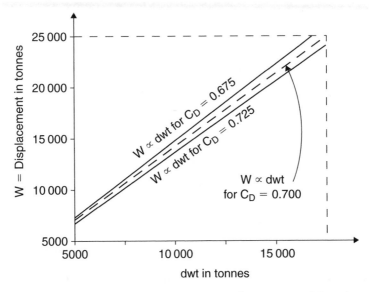

Fig. 1.5   W ∝ dwt for General Cargo ships for a range of C_D values.

As can be seen in Figure 1.3, a mean line through the plotted points gave the equation:

$$L = 5.32 \times dwt^{0.351} \text{ m}$$

Figures 1.4 and 1.5 show more relationships to assist the designer in fixing the Main Dimensions for a new General Cargo vessel.

When selecting LBP for equations (1) and (2), for most Merchant ships at SLWL, we will soon know if practical values have been inserted.

If $C_B > 1.000$ this is impossible!!

If $C_B < 0.500$ this is improbable!!

Worked example 1.6

Estimates for a 500 000 tonnes are being considered. Service speed is to be 16 kt operating on a maximum draft of 25.5 m with a C_D of 0.861.

Calculate the LBP, Breadth Mld, C_B, W and lwt if it is assumed that:

$$B = 0.24L - 28 \text{ m}   \text{ and }   C_B = 1.066 - V/(4 \times L^{0.5})$$

$$C_D = dwt/W   \text{ So }   W = dwt/C_D$$

Thus $W = 500\,000/0.861 = 580\,720$ tonnes

$lwt = W - dwt = 580\,720 - 500\,000 = 80\,720$ tonnes

$W = L \times B \times H \times C_B \times p$

$C_B = W/(L \times B \times H \times p)$

So $C_B = 580\,720/\{L \times (0.24L - 28) \times 25.5 \times 1.025\}$

$$C_B = 22\,218/\{L(0.24L - 28)\} \tag{1}$$

$$C_B = 1.066 - V/(4 \times L^{0.5})$$

$$C_B = 1.066 - 4/L^{0.5} \tag{2}$$

Now equation (1) = equation (2)

Substitute values for L of 380, 390 and 400 m. Draw graphs (as before) of L against C_B values. At the point of intersection,

$$L = 391\,m \quad \text{and} \quad C_B = 0.863$$

$$B = 0.24L - 28 = (0.24 \times 391) - 28 = 65.84\,m$$

$$H = 25.5\,m, \text{ as prescribed}$$

$$W = 391 \times 65.84 \times 25.5 \times 0.863 \times 1.025 = 580\,686 \text{ tonnes,}$$

which is very close to the previous estimate of 580 720 tonnes.

Depth Mld (D) for the new design

Again guidance can be given by careful selection of a basic ship or basic ships. The following approximations can be considered:

For Oil Tankers	H/D = 80% approximately
For General Cargo ships	H/D = 75% approximately
For liquified natural gas (LNG) and liquified petroleum gas (LPG) ships	H/D = 50% approximately

After obtaining draft H, simply transpose to obtain value of D. Freeboard (f) is the difference between these two values.

Freeboard (f) on Oil Tankers

It can be seen from the given H/D percentages that the summer freeboard for the General Cargo ships will be approximately 25%. For the Oil Tankers it is more likely to be nearer 20%.

Freeboard on Oil Tankers have *less freeboard* than General Cargo ships of similar length for several reasons, six of them being:

1. Smaller deck openings in the Upper Deck.
2. Greater sub-division by transverse and longitudinal bulkheads.
3. Density of cargo oil is less than grain cargo.
4. Much larger and better pumping arrangements on tankers to control any ingress of bilge water.
5. Permeability for an oil-filled tank is only about 5% compared to permeability of a grain cargo hold of 60–65%. Hence ingress of water in a bilged compartment will be much less.
6. Larger Transverse Metacentric Height (GM_T) values for an Oil Tanker, especially for modern wide shallow draft tanker designs.

Optimisation of the Main Dimensions and C_B

Early in the design stages, the Naval Architect may have to slightly increase the displacement. To achieve this, the question then arises, 'which parameter to increase, LBP, Breadth Mld, depth, draft or C_B?'

Increase of L

This is the most expensive way to increase the displacement. It increases the first cost mainly because of longitudinal strength considerations. However, and this has been proven with 'ship surgery', there will be a reduction in the power required within the engine room. An option to this would be that for a similar input of power, there would be an acceptable increase in speed.

Increase in B

Increases cost, but less proportionately than L. Facilitates an increase in depth by improving the transverse stability, i.e. the GM_T value. Increases power and cost within the machinery spaces.

Increases in Depth Mld and Draft Mld

These are the cheapest dimensions to increase. Strengthens ship to resist hogging and sagging motions. Reduces power required in the Engine Room.

Increase in C_B

This is the cheapest way to simultaneously increase the displacement and the deadweight. Increases the power required in the machinery spaces, especially for ships with high service speeds. Obviously, the fuller the hull-form the greater will be the running costs.

The Naval Architect must design the Main Dimensions for a new ship to correspond with the specified dwt. Mistakes have occurred. In most ship contracts there is a severe financial penalty clause for any deficiency in the final dwt value.

Questions

1. For a 'STAT 55' proposal it is known that: L/B is 6.23, B/H is 2.625, C_B is 0.805, C_D is 0.812, dwt is 55 000 tonnes. Calculate the LBP, Breadth Mld, W and lwt for this proposed design.
2. Define and list the components for: (a) lightweight, (b) deadweight, (c) load displacement, (d) block coefficient C_B, (e) deadweight coefficient C_D.
3. From a database, information for a new ship is as follows: C_D is 0.701, $B = (L/10) + 6.72$, dwt is 13 750 tonnes, service speed is 14.5 kt, Draft Mld is to be a maximum of 8.25 m. Estimate the LBP, Breadth Mld, C_B, and fully loaded displacement.

4 A 110 000 tonnes dwt tanker is 258 m LBP, 43 m Breadth Mld and 14.20 m Draft Mld. A new similar design of 120 000 tonnes is being considered. Using the geosim method, estimate the LBP, Breadth Mld and Draft Mld for the larger ship.

5 Three new standard General Cargo vessels are being considered. They are to have deadweights of 4500, 8500 and 12 500 tonnes respectively. Estimate (as a first approximation), the LBP for each of these ships.

6 A container ship is to have a service speed of 21.5 kt and an LBP of 180 m. Using two methods, estimate her C_B value at her Draft Mld.

Chapter 2

Preliminary estimates for group weights for a new ship

Section 1

Estimation of steel weight for a new ship

For every ship there is a 'balance of weights' table, an example of which is shown in Table 2.1. This shows the actual figures for a Shelter deck General Cargo vessel of 128 m length between perpendiculars (LBP).

Table 2.1 A balance of weights table (tonnes)

Steel weight	2800
Wood and Outfit weight	700
Hull weight	3500
Machinery weight	550
Lightweight	4050*
Deadweight	9050
Fully loaded weight	13 100

C_D = deadweight (dwt)/W = 9050/13 100 = 0.691.
*Maximum margin of error to be less than 2% of the lightweight or 4.5 × Tonnes per centimetre Immersion (TPC) at the Summer Loaded Waterline (SLWL).

The Naval Architect will always attempt to make the lightweight as low as possible without endangering the safety and strength of the new vessel. The Department of Transport (DfT) and International Maritime Organisation (IMO) keep a watchful eye on the safety standards whilst Lloyds are more concerned with the strength considerations. Other countries have equivalent Classification Societies.

Consideration of steel weight estimations

The main factors affecting the steel weight are:

Dimensions L, B, D, H   Block coefficient
Proportions L/B, B/H, L/H, etc.   Deckhouses

Length of superstructures	Mast-houses
Number of decks	Deck sheer
Number of bulkheads	Engine seatings

Net scantling weight: This is the steel weight that is actually ordered in by the shipyard. It is subjected to a rolling margin of −2.5% to +2.5% of the thickness of each plate.

Invoice weight: This is the steel purchased by the shipyard.

Net steel weight: This is the weight that ends up in the new ship. It takes into effect the wastage caused by plate preparation. The steel that ends up on the cutting floor can be 8–10% of the delivered plate. Figure 2.1 shows a nested plate with wastage material regions.

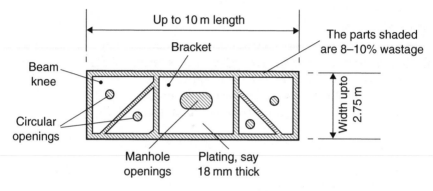

Fig. 2.1   A nested steel plate.

Methods for estimating steel weight in ships

There are several methods for obtaining the steel weight of a new design some of them being:

1. Cubic Number method
2. Weight per metre method
3. 'Slog-slog' method
4. Method of differences
5. Computational techniques.

Cubic Number method

This should only be used for preliminary or tentative estimates:

$$\text{Cubic No.} = \frac{L \times B \times D}{100}$$

where
L = LBP,
B = Breadth moulded (Br.Mld),
D = Depth Mld.

If in similar ships the Main Dimensions vary as L, then the weights will vary as L cubed. This is only true if B and D vary in the same proportion as L. Thickness in scantlings will vary in the same proportion.

This will seldom occur. Thus considerable error can result if the Cubic No. is directly applied. It is more efficient to obtain proportional dimensions for the new design using the Cubic No. and then adjusting for differences in the values of B and D. These adjustments are explained in detail later in this chapter.

Worked example 2.1

A basic ship is $121.95\,\text{m} \times 16.46\,\text{m} \times 9.15\,\text{m}$ Depth Mld with a total steel weight of W_s. A new similar design is $131.1\,\text{m} \times 17.07\,\text{m} \times 10.06\,\text{m}$ Depth Mld. Show the method for obtaining the steel weight for the $131.1\,\text{m}$ ship.

	L	B	D	
Basic ship	121.95	16.46	9.15	
Basic ship $\times L_2/L_1$	131.10	17.70	9.83	(1)
New design dimensions	131.10	17.07	10.06	(2)
Equation (2) − (1)	= zero	−0.63	+0.23	

Thus the steel weight for the new design = $W_s \times (131.10/121.95)^3$ but with a deduction for 0.63 m of breadth and an addition of 0.23 m for depth. The fourth method later shows how the adjustments are then made for this deduction and this addition.

Weight per metre run method

In this method it is necessary to have the midship sections of the basic ship and for the new ship. Calculations are made to obtain the weight per metre run at amidships for both ships. In these calculations only longitudinal plating and longitudinal stiffening are considered. Intercostal steel structures are ignored.

Worked example 2.2

Weight per metre run (W_b) for a basic ship is 13.12 tonnes/m. LBP is 118 m and steel weight is 2350 tonnes. From the preliminary midship section for the new design, the weight per metre run (W_d) is 13.76 tonnes/m. LBP is 124 m. Estimate the steel weight for the new design.

Let steel weight for the basic ship = W_b

Let steel weight for the new design = W_d

Then
$$W_d = W_b \times \left(\frac{W_d}{W_b}\right) \times \left(\frac{L_2}{L_1}\right)$$

$$= 2350 \times \frac{13.76}{13.12} \times \frac{124}{118}$$

$$= 1.102 \times 2350$$

$$= 2590 \text{ tonnes}$$

Note that this is only a first approximation and must always be treated as such. There are certain assumptions with this method. One is that the various parts of the two ships have the same proportions to each other throughout their lengths as they do at their respective amidships.

It is also assumed that the vessels have proportionate sheer, extent of decks, deck openings, etc. Furthermore, it is assumed that the graduation of scantlings towards the ends on each vessel is equally proportional to steel thicknesses at amidships.

Modifications or corrections for non-compliance with these assumptions must be made. Differences in the general arrangements of both ships must also be taken into account.

Because of these assumptions, adjustments will then be made to the first estimate of 2590 tonnes calculated in Worked example 2.2.

The 'slog-slog' method

This method is used where a basic ship is not available. It requires a preliminary set of steel plans for the new design. Length, breadth and thickness of the steel plates and stiffeners are multiplied together, and then added to give a total volume of steel. Any openings in the steel have to be allowed for and deducted from this volume.

By bringing in the specific gravity for steel of about 7.85, the volume can be changed to steel weight. Being very repetitive in nature it is very tedious. It can take a long time to obtain the final steel weight. This is why it is known as the 'slog-slog' method!!

Method of differences

In this method, dimensional correction is made for length, breadth and depth after comparisons have been made between the new design and a selected basic ship.

Feedback from ships already built has shown that the steel weight in tonnes/m run for length, breadth and depth are as follows:

- 85% is affected by length of a ship,
- 55% is affected by the breadth of a ship,
- 30% is affected by the depth of a ship,
- 45% is affected by the depth of a ship for Oil Tankers only.

The percentages take into account end curvature of vessels and curvature below say the Upper Deck level.

Worked example 2.3

A General Cargo ship is 122 m × 16.45 m × 9.20 m Depth Mld. She has a finished steel weight of 2700 tonnes. The new ship has preliminary dimensions of 131 m × 17.08 m × 10.10 m Depth Mld. Estimate the steel weight for the new design after correcting for the Main Dimensions only.

For the basic ship:

Rate along the length $= 85\% \times (2700/122)$ $= 18.81$ tonnes/m run

Rate across the breadth = 55% × (2700/16.45) = 90.27 tonnes/m run

Rate down the depth   = 30% × (2700/9.20) = 88.04 tonnes/m run

	L	B	D
Basic ship	122	16.45	9.20
New design	131	17.08	10.10
Differences	+9	+0.63	+0.90
Rates in tonnes/m run	18.81	90.27	88.04
Modifications	+169	+57	+79 = +305 tonnes

So, new design's steel weight = basic steel weight + modifications

$$= 2700 + 305$$
$$= 3005 \text{ tonnes} \quad \text{after modifying for Main Dimensions only!!}$$

Note how the three rates in tonnes/m for the basic ship, are also used for the new design. It should also be realised that any or all of the three modifications can be positive, zero or indeed negative.

Worked example 2.4

A basic General Cargo ship is 135 m × 18.53 m × 10.0 m Depth Mld with a finished steel weight of 3470 tonnes. A new design is 136.8 m × 18.36 m × 9.8 m Depth Mld. Estimate the steel weight for the new design after modifying for Main Dimensions only.

For the basic ship,

Rate along the length   = 85% × 3470/135 = 21.85 tonnes/m run

Rate across the breadth = 55% × 3470/18.53 = 103.0 tonnes/m run

Rate down the depth   = 30% × 3470/10   = 104.10 tonnes/m run

	L	B	D
Basic ship	135.0	18.53	10.0
New design	136.8	18.36	9.8
Differences	+1.8	−0.17	−0.2
Rates in tonnes/m run	21.85	103.0	104.1
Modifications	+39	−18	−21 = zero

So, new design's steel weight = basic steel weight + modifications

$$= 3470 + \text{zero}$$
$$= 3470 \text{ tonnes} \quad \text{similar to basic ship steel weight!!}$$

After modifying for dimensions only, it is necessary to modify further, for further differences in the steel structures between the basic ship and the new design. This will be as follows.

Modification for C_B

The correction is ±½% for each 0.010 change in the C_B at the Summer Loaded Waterline (SLWL). Reconsider Worked example 2.3 where the steel for the new design after correcting for dimensions was 3005 tonnes. Suppose the respective C_B values at their respective SLWLs were 0.725 for the basic ship and 0.740 for the new design.

$$C_B \text{ correction} = \frac{0.740 - 0.725}{0.010} \times \left(+\frac{1}{2}\%\right) \times 3005 = +23 \text{ tonnes}$$

Scantling correction

This can be taken as a fraction of each of the dimensional corrections. It is in effect a modification for differences in the proportions of the Main Dimensions. Feedback from ships already built suggest that these scantling corrections should be:

$(1/3) \times$ Length correction in tonnes
$(1/4) \times$ Breadth correction in tonnes
$(1/2) \times$ Depth correction in tonnes

Reconsider the Worked example 2.3 where the modifications were +169, +57 and +79 tonnes. Then:

$$\text{Scantling correction} = \frac{169}{3} + \frac{57}{4} + \frac{79}{2} = +110 \text{ tonnes}$$

Deck sheer correction

This correction is obtained by first calculating the mean deck sheer for basic ship and new design. Calculate the difference in these answers and then multiply it by the depth correction rate in tonnes/m run.

$$\text{Mean deck sheer for both ships} = \frac{\text{Sheer aft} + \text{Sheer for'd}}{6}$$

Table 2.2 Table of corrections or modifications to basic ship's steel weight of 2700 tonnes

Item	Positive	Negative	Item	Positive	Negative
Dimensions	305	–	Watertight bulkheads		–
C_B correction	23	–	Non-watertight bulkheads	–	
Scantlings	110	–	Deep tanks	–	
Sheer correction	13	–	Oil fuel bunkers		–
Bulwarks		–	Machinery casings		–
Poop deck	–		Shaft tunnel		–
Bridge deck		–	Double bottom	–	
Boat deck	–		Minor decks	–	
Wheelhouse top	–		Miscellaneous items	–	
Total	**A**	**B**		**C**	**D**

The finished steel weight for the new design will then = 2700 + **A** − **B** + **C** − **D** tonnes.

Assume for the first example that the basic ship has aft sheer of 1.27 m and for'd sheer of 2.75 m with the new design having 1.38 m aft sheer and 3.5 m for'd sheer. Calculate the sheer correction in tonnes.

$$\text{Sheer correction} = \left(\frac{1.38 + 3.50}{6} - \frac{1.27 + 2.75}{6} \right) \times 88.04 = +13 \text{ tonnes}$$

There are other modifications to consider. These are shown in Table 2.2. On each occasion the differences are examined between the basic ship and the new design and the modification to the steel weight tabulated.

Computational techniques

Many formulae have been suggested by researchers for estimating the finished steel weight. Three of them have been J.M. Murray, I. Buxton and S. Sato. They were derived after keying in and plotting a lot of detailed total steel weights from ships already built and in service. They were all for Supertankers.

$$W_{ST} = 26.6 \times 10^{-3} \times L^{1.65} \frac{(B + D + H/2)(0.5C_B + 0.4)}{0.8} \text{ tonnes}$$

J.M. Murray (1964)

$$W_{ST} = a_1 \times (L^{1.8} \times B^{0.6} \times D^{0.4})(0.5C_B + 0.4) \text{ tonnes} \qquad \text{I. Buxton (1964)}$$

$$W_{ST} = \left(\frac{C_B}{0.8} \right)^{1/3} \times 10^{-5} \left\{ \left(5.11 \times L^{3.3} \times \frac{B}{D} \right) + (2.56 \times L^2 \times (B + D)^2) \right\} \text{ tonnes}$$

S. Sato (1967)

where:

L = LBP,
B = Br.Mld,
D = Depth Mld
H = Draft Mld
C_B = block coefficient at SLWL
a_1 = Buxton's coefficient to obtain units of tons (or tonnes).

These formulae serve only to give only *first* approximations to the steel weight. As ship main proportions have changed over the years and as high tensile steel became more used in these Supertankers then the coefficients will also have changed with time. Treat these computer derived formulae with caution, and certainly only as a first guidance to the finished steel weight.

Of the five methods discussed, it is suggested the best one is the 'Method of differences.'

Prefabrication techniques – a short note

Having discussed at length the calculations for predicting the steel weight for a new ship, it is now appropriate to briefly look at the design assembly line for this steel weight in a shipyard. Figure 2.2 shows the planned route for the steel from the stockyard, through the various sheds and finally to be fitted onto the ship on her berth.

The advantages of these prefabrication methods are:

1. It is much quicker to build and launch the ship. For some General Cargo ships, it takes only 3 weeks from the time of laying of the first keel plate, to the time that the vessel is launched.
2. Because of reduced labour costs, it is thus cheaper to build a ship.
3. Much work can be completed under cover and thus less time lost to bad weather conditions.
4. More automation can be employed for cutting and welding of plating. With modern systems computer tapes (CADAM) eliminate even the need to mark the plates prior to cutting them.
5. More down-hand welding can be performed. This is achieved by turning the units over in a prefabrication shed. Consequently, faster and more efficient jointing is achieved.
6. There is a less cluster of workers stopping one another from working whilst one operative is waiting for another to finish a job before starting on their particular task.
7. It is much easier to modify a curved plate in a prefabrication shed than at an open air ship's berth.

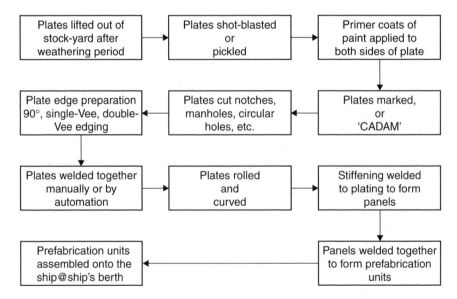

Fig. 2.2   Prefabrication method for the steel work of a new ship.

Section 2

Wood and Outfit weight

This weight generally includes everything in the hull weight except the net steel weight. Many weights have to obtained separately. In certain cases the finished weight can be obtained from the sub-contractors. They could be supplying equipment such as winches, windlass, lifeboats, fridge machinery, galley equipment, hold and tween deck insulation, navigation instruments, etc.

Most of the Wood and Outfit (W&O) weight will be generally situated within the accommodation spaces. There are two popular methods for obtaining the final (W&O) weight for a new ship.

Method 1: The coefficient procedure

This method requires calculating a coefficient 'α_B' for a basic ship and then using the same coefficient for the new similar design.

$$\alpha_B = \frac{\text{W\&O weight for basic ship} \times 100}{L_B \times B_B}$$

$$\text{W\&O weight for new design} = \alpha_B \times \frac{L_D \times B_D}{100} \text{ tonnes}$$

The coefficient 'α' depends upon the standard of accommodation, number of crew, refrigerated stores, etc. For a General Cargo ship or Oil Tanker the value of will be of the order of 20–30. It is very important to take care with the selection of the basic ship when comparing her with the new design. They must be similar in type, and close in size, speed and power.

Method 2: Proportional procedure

A second method is to assume that part of the W&O weight is affected by the dimensions of L and B. How much depends on the ship-type being considered.

For new General Cargo ships:

$$\text{W\&O weight} = \left(\frac{\text{W\&O weight}}{2}\right)_B + \left(\frac{\text{W\&O weight}}{2}\right)_B$$
$$\times \left(\frac{L_D \times B_D}{L_B \times B_B}\right) \text{ tonnes}$$

For new Oil Tankers:

$$\text{W\&O weight} = \frac{2}{3}(\text{W\&O weight})_B + \frac{1}{3}(\text{W\&O weight})_B$$
$$\times \left(\frac{L_D \times B_D}{L_B \times B_B}\right) \text{ tonnes}$$

Worked example 2.5

A basic General Cargo ship is 134 m LBP × 18.12 m Br. Mld with a final W&O weight of 700 tonnes. A new similar ship has an LBP of 138.5 m and a Br. Mld of 18.70 m. Estimate the W&O coefficient 'α_B' and the new W&O weight for the new design.

Method 1

$$\alpha_B = \frac{\text{W\&O weight for basic ship} \times 100}{L_B \times B_B}$$

$$= \frac{700 \times 100}{134 \times 18.12} = 28.83$$

$$\text{W\&O weight for new design} = \alpha_B \times \frac{L_D \times B_D}{100} \text{ tonnes}$$

$$= \frac{28.83 \times 138.5 \times 18.7}{100}$$

$$= 746 \text{ tonnes}$$

Method 2

$$\text{W\&O weight} = \left(\frac{\text{W\&O weight}}{2}\right)_B + \left(\frac{\text{W\&O weight}}{2}\right)_B$$
$$\times \left(\frac{L_D \times B_D}{L_B \times B_B}\right) \text{ tonnes}$$

$$\text{W\&O weight for new design} = \frac{700}{2} + \frac{700}{2} \times \frac{138.5 \times 18.70}{134 \times 18.12}$$

$$= 723 \text{ tonnes}$$

These values of 746 and 723 tonnes are first estimates only and must always be treated as such. Method 1 gives perhaps the better prediction because it is based on data from one very similar ship. Method 2 is a format based on average feedback from several ships. A third option would be to take a mean value of the two answers, thereby giving a value of 735 tonnes as the first estimation.

In 1984, feasibility studies carried out by British Shipbuilders Ltd produced a multipurpose vessel (MP17) with the following data:

- 17 000 tonnes dwt.
- 17 tones/day for the oil fuel consumption.
- 17 person complement.

This design obviously requires fewer cabins, fewer communal rooms, less heating, lighting and ventilation, etc. Hence, the W&O weight and coefficient 'α_B' will have lower values.

In the 1950s, it was 45–50 in a crew. In August 2003, it is usual to have crews of 18–24 on tankers and General Cargo ships. As a consequence, 'α_B' will be at the lower end of the previously quoted range of 20–30.

After using Methods 1 and 2, further modifications need to be made for any differences in the W&O arrangements between the basic ship and the new design. A tabulated statement bringing all these differences together as a total, in conjunction with the first estimate, will give the final W&O weight for the new design.

Non-ferrous metals

Non-ferrous metals may be included in the final W&O weight. The use of these metals is extensive throughout a ship. They include:

- *Aluminium alloys*: Fitted in navigation spaces because of their non-magnetic characteristics. Lighter in weight than steel. Not as corrosive as steel. Not so brittle as steel at low temperatures. Fitted in cargo tanks on liquefied natural gas (LNG) and liquefied petroleum gas (LPG) ships.
- *Brass*: Used for small items such as sidelights, handrails, sounding pipe caps, plus rudder and propeller bearings.
- *Copper*: Used mainly for steam pipes. Copper is a soft pure metal that is malleable and ductile.
- *Zinc*: Used as sacrificial anodes around a ship's rudder and sternframe. The zinc acts as an anode. In time due to galvanic action the zinc is eaten away and the steelwork around the propeller's aperture remains relatively unharmed.
- *Lead*: This is a soft heavy pure metal often used for service piping.
- *Manganese bronze*: Used in the construction of propellers. Note that this item of weight will be included in the machinery weight total for a new ship.

Use of plastics for Merchant ships

Since 1980, plastics have been used more and more for ship structures. They have for some structures replaced steel, wood or aluminium. The main advantages of fitting plastics on ships can be one or more in the following list:

- Weight saving
- Non-corrosive
- Non-magnetic
- Rot-resistant
- Abrasion resistant
- Easy maintenance/renewal
- Ability to tailor
- Smooth frictional characteristics
- Chemical resistant
- Heat/electrical insulator
- Moisture non-retainer
- Decorative – aesthetically pleasing
- Transparency qualities
- Adhesive properties.

Fibreglass for example does not rot, warp or twist. This makes it particularly advantageous over wood. To be used effectively on ships, thermoplastics and thermo-setting plastics must offer certain basic qualities. For example:

- adequate strength,
- resistance to corrosion (oxidation and galvanic),
- ability to be worked into structural shapes,
- least weight, but with adequate strength,

- lower first costs,
- low fire risk.

Plastics have been used on ships for the following structures:

- bulkhead facings – accommodation blocks, replacing paint,
- cabin furniture – replacing wood,
- deck awnings – replacing canvas or aluminium,
- lifeboats – replacing wood, steel, or aluminium,
- sidelights and windows – replacing steel or brass,
- cold water piping – replacing steel,
- deck floor coverings in accommodation and navigation spaces,
- electrical fittings such as cable trays,
- mooring lines – replacing hemp,
- insulation in reefer ships – replacing cork,
- tank top ceilings – replacing wood,
- sounding and ullage pipes – replacing steel,
- superstructures on small luxury craft – replacing steel or aluminium.

A lot of these structures will be manufactured outside of the shipyard. They will be made by sub-contractors. They must supply the shipyard with a written note of the weight(s) of their product for inclusion in the 'balance of weights' table.

Plastics offer the Naval Architect possibilities of a lowering of the new ship's lightweight but should always be with the proviso that they do not reduce the seaworthiness aimed for by the design team.

Section 3

Estimations of machinery weight

The total machinery weight includes:

- the main engine,
- the auxiliary machinery,
- propeller,
- propeller shaft,
- engine spares.

Method 1: The rate procedure

One method is to use the machinery power in kW and divide it by the total machinery weight in tonnes. This gives a rate measured in kW/tonnes and is used for both the basic ship and the new design.

Worked example 2.6

Data for a basic ship is as follows:

Brake power $P_B = 5250\,kW$,   displacement $W = 13\,500$ tonnes,
service speed $= 16\,kt$,   total machinery weight $= 680$ tonnes.

A new similar design is being considered. She has a displacement of 14 100 tonnes with a service speed of 16.25 kt. Estimate the total machinery weight for the new design.

For the basic ship's machinery, $\text{Rate} = \dfrac{\text{Power}}{\text{Weight}} = \dfrac{5250}{680} = 7.72 \text{ kW/tonnes}$

Note: the higher this rate is the better and more efficient is the ship's machinery.

For similar ships, we can use the same rate in kW/tonnes and also the same Admiralty coefficient (A_C), where:

$$A_C = \frac{W^{2/3} \times V^3}{P}$$ 350–600 for Merchant ships, the higher values being for the better-designed ships

where:

W = ship's displacement in tonnes,
V = ship's service speed in kt with $V < 20$ kt,
P = power in kW,
  = P_B for brake power in Diesel machinery,
  = P_S for shaft power in Steam Turbine machinery.

Caution: If V equals 20 kt or more then use V^4 instead of V^3. This will assist in making more accurate comparisons when dealing with similar higher speed vessels.

For this worked example, now calculate the brake power P_B for the new design:

$$A_C(\text{basic}) = A_C(\text{design})$$

$$\frac{13\,500^{2/3} \times 16^3}{5250} = \frac{14\,100^{2/3} \times 16.25^3}{P_B(\text{design})}$$

$$442 = \frac{2\,502\,812}{P_B(\text{design})}$$

$$P_B(\text{design}) = \frac{2\,502\,812}{442} = 5562 \text{ kW}$$

For the new design, total machinery weight $= \dfrac{\text{New power (kW)}}{\text{Rate (kW/tonnes)}}$

$$= \frac{5562}{7.72} = 720 \text{ tonnes}$$

This value represents a first prediction for the machinery weight. Further modification must then be made for any differences between the basic ship and the new design's arrangement of machinery installation. This will finally give what is known as the 'all-up' machinery weight.

Having obtained the total machinery weight it is then possible to predict the weight of the main engine. The following approximations may be used:

For Diesel machinery, $\dfrac{\text{Main engine weight}}{\text{All-up weight}} = \dfrac{3}{7}$ approximately

For Steam Turbines, $\dfrac{\text{Main engine weight}}{\text{All-up weight}} = \dfrac{1}{7}$ approximately

For Peilstick Diesel machinery,   Main engine weight = 1/4 times a Doxford or a Sulzer main engine (approximately).

Method 2: Use of empirical formulae

Several researchers have produced empirical formulae for predicting the 'all-up' machinery weight (M_W). They offer a first attempt, when knowing only the brake power P_B or the shaft power P_S. Feedback from existing ships have shown that:

For Diesel machinery, $M_W = 0.075P_B + 300$ tonnes   C.B. Barrass

where $P_B = 5500$–$13\,000$ kW   See Figure 2.3

For Steam Turbines, $M_W = 0.045P_S + 500$ tonnes   C.B. Barrass

where $P_S = 13\,000$–$24\,250$ kW   See Figure 2.4

For Steam Turbines, $M_W = 10.2 \times (P_S)^{0.5}$ tonnes   D.G.H. Watson

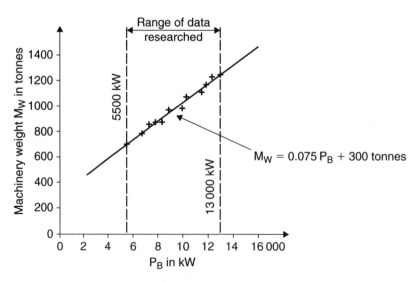

Fig. 2.3 $M_W \propto P_B$ for diesel machinery.

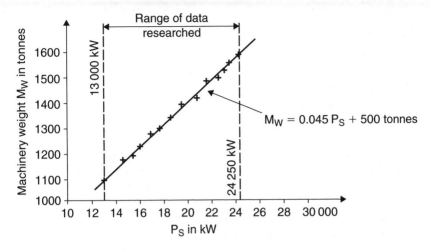

Fig. 2.4 $M_W \propto P_S$ for Steam Turbine machinery.

Worked example 2.7

A ship of 9500 tonnes dwt has power at the thrust block of 5000 kW (either P_B or P_S). Estimate the total machinery weight when diesel machinery is fitted or when Steam Turbine machinery is installed in this ship:

For Diesel machinery, $M_W = 0.075 P_B + 300$ tonnes   C.B. Barrass

$$= (0.075 \times 5000) + 300 = 675 \text{ tonnes}$$

For Steam Turbines, $M_W = 0.045 P_S + 500$ tonnes   C.B. Barrass

$$= (0.045 \times 5000) + 500 = 725 \text{ tonnes}$$

For the Diesel machinery; installed on single screw ships, propeller revolutions were 120 rpm, with a service speed of about 16 kt. They were of Doxford or Sulzer design.

For the Steam Turbine machinery; installed on single screw ships, propeller revolutions were 80–85 rpm, with service speeds 15–15.5 kt. They were of AEI or Stal-Laval design.

For Steam Turbines, $M_W = 10.2 \times (P_S)^{0.5}$ tonnes   as per Watson

$$M_W = 10.2 \times 5000^{0.5} = 721 \text{ tonnes}$$
(close to value, via Barrass formula).

Machinery weight adjustments

1. If the machinery weight is all-aft (as on Oil Tankers) instead of being located at amidships, then reduce the total 'all-up' weight by 5%. This allows for reduction in length of shafting and shaft supports.
2. If the vessel is twin screw then add about 10%. This allows for additional propeller shaft structures.
3. If the machinery is heavily electrically loading, then add 5–12%.

Questions

Section 1

1 List the components that make up a 'balance of weights' table for a ship.
2 Define the following steel weight terms:
   (a) Net scantling steel weight,
   (b) Invoiced steel weight,
   (c) Net steel weight,
   (d) A nested plate.
3 List the factors that affect the steel weight for a basic ship or a new design.
4 A basic ship has an LBP of 121 m with a midship rate of 12 tonnes/m run and a finished steel weight of 2750 tonnes. Estimate, as a first approximation, the steel weight for a new similar design that has an LBP of 125 m and a midship rate of 12.25 tonnes/m run.
5 The following information is known for a basic General Cargo ship and a similar new design:

Item	Basic ship	New design
LBP (m)	140	145
Br. Mld (m)	19.5	20.5
Depth Mld (m)	12.6	12.3
C_B at SLWL	0.726	0.735
Aft deck sheer (m)	1.52	1.43
For'd deck sheer (m)	3.20	2.94
Residual steel additions (tonnes)	–	+39
Total finished steel weight (tonnes)	4035	xxxx

Estimate the steel weight for the new design after modifications have been made to the basic ship's steel weight for Main Dimensions, C_B, proportions, sheer and residual additions.

6 Sketch a diagram of a modern prefabrication assembly line for the steel work for a new ship. List five advantages of building ships when using prefabrication methods.

Section 2

1 List the items generally included in the W&O weight for a new ship.
2 List reasons why the W&O weight is less today compared to say 15 years ago.
3 Why are plastics fitted on ships? Suggest for which ship structures, plastics may be used?
4 Name four non-ferrous metals and suggest whereabouts on a ship they may be fitted.
5 (a) Using the table of data, estimate the W&O weight for the new General Cargo ship by two methods for correcting for Main Dimensions only.

Vessel	LBP (m)	Br. Mld (m)	W&O weight
Basic ship	137.5	19.75	736 tonnes
New design	140.5	19.95	xxxx

(b) Give reasoning why one method should give a slightly more accurate prediction.

Section 3

1 List the components that make up the 'all-up' machinery weight.
2 A new ship has a displacement of 19 500 tonnes, a service speed of 14.7 kt and a brake power of 4950 kW. Calculate her admiralty coefficient (A_C).
3 A vessel has a power measured at the thrust block of 13 000 kW. Estimate the total machinery weight if:
(a) Diesel machinery was fitted,
(b) Steam Turbine machinery was installed.
4 Data for a selected basic ship with Diesel machinery is as follows: P_B = 4600 kW,   W = 15 272 tonnes,   V = 15.50 kt,   machinery   weight = 663 tonnes. A new similar design has: W = 14 733 tonnes, V = 15.25 kt. Estimate the machinery weight for the new design by two methods.
5 If the 'all-up' machinery weight for a ship is 560 tonnes, estimate approximately the weight of the main engine unit if:
(a) Diesel machinery is installed.
(b) Steam Turbine machinery is fitted.

Chapter 3

Preliminary capacities for a new ship

It is usual when dealing with ship capacities to consider:

Moulded Capacity.
Grain Capacity.
Bale Capacity.
Insulated volume.

- *Moulded Capacity*: This is the internal volume of a compartment, without taking into account stiffeners, frames, brackets, beams, girders, etc.

- *Grain Capacity*: This is the Moulded Capacity minus the volume taken up by the stiffeners, frames, brackets, beams, girders, etc. This stiffening is of the order of 1.5% of the Moulded Capacity. Hence:

  Grain Capacity = 98.5% × Moulded Capacity in m^3 approximately

- *Bale Capacity*: This is the volume measured to the inside of frames, to the underside of beams and to the top of the Tank Top ceiling. It about 10% less than the Grain Capacity. Hence:

  Bale Capacity = 90% × Grain Capacity in m^3 approximately

- *Insulated volume*: This is a volume that takes into account the insulation built into a compartment. Usually fitted on reefer ships. Thickness of insulation can range from being 200 to 350 mm. It is about 25% of the Moulded Capacity. Hence:

  Insulated capacity = 75% × Moulded Capacity in m^3 approximately

Worked example 3.1
For a vessel the Moulded Capacity is 20 000 m^3. Estimate the approximate corresponding grain, bale and insulated capacities.

   Grain Capacity = 98.5% × Moulded Capacity in m^3 approximately

$$= 98.5\% \times 20\,000 = 19\,700\,\text{m}^3$$

$$\text{Bale Capacity} = 90\% \times \text{Grain Capacity in m}^3 \text{ approximately}$$

$$= 90\% \times 19\,700 = 17\,730\,\text{m}^3$$

$$\text{Insulated capacity} = 75\% \times \text{Moulded Capacity in m}^3 \text{ approximately}$$

$$= 75\% \times 20\,000 = 15\,000\,\text{m}^3$$

Detailed estimation of the Grain Capacity

Consider first of all the total Grain Capacity extending from the Fore Peak bulkhead to the Aft Peak bulkhead, above the Tank Top extending to the uppermost continuous deck.

To this capacity add the volumes of the none-cargo spaces like access trunking, machinery spaces, etc. Assume for the selected basic ship that these totalled together gave a grand total of 'G_B.' To obtain the equivalent value for a new similar design 'G_D' the following formulae are used:

$$G_D = G_D \left(\frac{L_D \times B_D \times {'D_D'} \times C_B @ \text{SLWL}_D}{L_B \times B_B \times {'D_B'} \times C_B @ \text{SLWL}_B} \right) \text{m}^3$$

where:

G_D and G_B are measured in cubic metres,

$\quad$ L = length between perpendiculars (LBP) in metres,

$\quad$ B = Breadth moulded (Br. Mld) in metres,

$\quad$ C_B = block coefficient,

SLWL = Summer Loaded Waterline (Draft Mld in metres).

$$'D' = \text{Depth Mld} + \frac{\text{Camber}}{2} + \frac{\text{Sheer aft} + \text{Sheer for'd}}{6}$$
$$- \text{Tank Top height} - \text{Tank Top ceiling}$$

$$\frac{\text{Camber}}{2} = \text{Mean camber of the uppermost continuous deck}$$

$$\frac{\text{Sheer aft} + \text{Sheer for'd}}{6} = \text{Mean sheer of the uppermost continuous deck}$$

'D' is, in effect, the depth of the ship that is containing grain cargo.

When G_D has been obtained, all none-cargo spaces below the uppermost continuous deck must be deducted and any additional cargo capacity above the deck added in. For example, this additional capacity may be in the hatch coamings, or in the no. 1 Forecastle tween decks. Volume of hatch coamings will, in practice, be about ½% of the Grain Capacity for this type of ship. The final total will give the final value of the Grain Capacity for the new ship.

Worked example 3.2

For a basic ship and a new similar design, the following particulars are known:

Item	Basic ship	New design
LBP (m)	134.0	137.0
Br. Mld (m)	18.50	19.50
Depth Mld (m)	12.00	12.20
Grain Capacity (m³)	17600	–
Tank Top (m)	1.25	1.40
C_B@ SLWL	0.760	0.745
Deck sheer for'd (m)	2.52	3.20
Deck sheer aft (m)	1.20	1.46
Deck camber (m)	0.38	0.46
Tank ceiling (m)	0.06	0.06
None-cargo spaces (m³)	3700	4490

Estimate the final grain and bale capacities for this new design.

For the basic ship and the new design:

$$'D' = \text{Depth Mld} + \frac{\text{Camber}}{2} + \frac{\text{Sheer aft} + \text{Sheer for'd}}{6}$$
$$- \text{Tank Top height} - \text{Tank Top ceiling}$$

$$'D_B' = 12.00 + \frac{0.38}{2} + \frac{2.58 + 1.20}{6} - 1.25 - 0.06 = 11.50 \text{ m}$$

$$'D_D' = 12.20 + \frac{0.46}{2} + \frac{3.20 + 1.46}{6} - 1.40 - 0.06 = 11.75 \text{ m}$$

$$G_B = \text{Grain Capacity} + \text{None-cargo spaces} = 17\,600 + 3700$$
$$= 21\,300 \text{ m}^3$$

$$G_D = G_B \left(\frac{L_D \times B_D \times \, 'D_D' \times C_B@\text{SLWL}_D}{L_B \times B_B \times \, 'D_B' \times C_B@\text{SLWL}_B} \right) \text{m}^3$$

So

$$G_D = 21\,300 \left(\frac{137 \times 19.5 \times 11.75 \times 0.745}{134 \times 18.5 \times 11.5 \times 0.760} \right)$$

$$G_D = 22\,990 \text{ m}^3$$

This value must now be adjusted, by a deduction for the none-cargo spaces in the new design.

$$\text{Final Grain Capacity} = 22\,990 - 4490 = 18\,500 \text{ m}^3 \quad \text{for the new design}$$

$$\text{Bale Capacity} = 90\% \times \text{Grain Capacity} = 90\% \times 18\,500$$

$$\text{Bale Capacity} = 16\,650 \text{ m}^3$$

Note: In some text books, C_B values are calculated at 85% of the Depth Mld for both the basic ship and the new design. At most, they are going to be only about 1.5% above those at each SLWL respectively (see Table 3.1).

When they are divided one by the other as shown in previous example, the error is so small as to be negligible. Hence, this author prefers to use C_B values at SLWL instead of at 85% SLWL. What's more, they are so much more readily available in the post preliminary Main Dimensions stages of the new design.

Table 3.1 C_B values @ SLWL and @85% Depth Mld for several ship types

Ship type	C_B@SLWL	C_B@ 85% Depth Mld
VLCCs	0.825	0.837
Oil Tankers	0.800	0.812
Large Bulk Carriers	0.825	0.837
Small Bulk Carriers	0.775	0.787
General Cargo ships	0.700	0.711

VLCCs: Very Large Crude Carriers.

Cargo oil capacity for Oil Tankers

The total volume over the cargo network of tanks in an Oil Tanker may be estimated by the following expression:

$$V_t = L_t \times B \times D_t \times C_B \times 1.16\,m^3$$

where:

V_t = Cargo oil tanks + water tanks capacity,
L_t = length over Cargo tanks network,
C_B = block coefficient@SLWL,
D_t = depth of Cargo tanks at amidships*,
1.16 = a hull-form modification coefficient for tankers, based on feedback of ships built and service. It is linked with the amount of parallel body designed into this type of vessel. For some Tankers the parallel body can be 65% of the ship's length LBP.

Worked example 3.3

An Oil Tanker has the following information:

LBP = 264 m, Br. Mld = 40.7 m, Depth Mld = 22.00 m, SLWL = 16.75 m, W = 151 000 tonnes, C_B@SLWL = 0.820, water ballast tanks within the cargo tank network = 15 000 m³, Fore Peak tank = 10 m long, Aft Peak tank = 10 m

---

*In the last decade, on new Oil Tankers, double bottoms have been fitted below the main cargo network of oil tanks. This must be accounted for. Consider the Worked example 3.3.

long, Deep tanks for'd = 10 m long, Engine Room = 31 m long, double bottom volume under main network of tanks = 16 000 m³, Allow 2% expansion due to heat in the Cargo oil tanks.

Calculate the Cargo Oil capacity for this Tanker.

$$V_t = L_t \times B \times D_t \times C_B \times 1.16 \, m^3$$

L_t = LBP − Fore Peak tank − Aft Peak tank − Deep tanks − Engine Room

$$= 264 - 10 - 10 - 10 - 31 = 203 \, m.$$

D_t = Depth Mld at this stage. However, the double-bottom volume under the Cargo tanks network will later be deducted.

$$V_t = 203 \times 40.7 \times 22 \times 0.820 \times 1.16 = \quad 172\,896 \, m^3$$

$$-\text{water ballast capacity (as given)} = \quad -15\,000 \, m^3$$

$$-\text{double-bottom capacity (as given)} = \quad \underline{-16\,000 \, m^3}$$

$$\text{Cargo oil tank capacity} = \quad 141\,896 \, m^3$$

$$-2\% \text{ expansion due to heat (as given)} = \quad \underline{-2838 \, m^3}$$

$$\text{Final Cargo oil tank capacity} = \quad \underline{139\,058 \, m^3}$$

So Final Cargo oil capacity = 139 058 m³

Capacity estimate for a Bulk Carrier

The capacity can be worked out in similar fashion the previous Oil Tanker problem. The following example shows the method of working.

Worked example 3.4

The dimensions for a proposed 60 000 tonnes deadweight Bulk Carrier have been estimated to be:

LBP = 235 m, Br. Mld = 31.5 m, Depth Mld = 18.0 m, C_B@SLWL = 0.827, length of Fore Peak tank = 11.75 m, length of Aft Peak tank = 8.25 m, Machinery Space length = 30.00 m, Upper Deck camber = 0.60 m, Upper Deck sheer = zero, hull-form coefficient = 1.19, Tank Top = 2 m above base.

Calculate the total volume; within the hold length, under the Upper Deck and above the Tank Top; before any adjustments have been made for topside tanks and side hoppers.

Let total length of holds = L_h

L_h = LBP − Fore Peak tank − Aft Peak tank − Machinery space length

$$= 235 - 11.75 - 8.25 - 30.00 = 185 \, m$$

$$D_h = \text{Depth Mld} - \frac{\text{Camber}}{2} - \frac{\text{Sheer aft} + \text{Sheer for'd}}{6} - \text{Tank Top height}$$

$$D_h = 18.00 - \frac{0.60}{2} - 0 - 2.00 = 16.30 \, m$$

Total required volume $= L_h \times B \times D_h \times C_B$@SLWL $\times$ hull-form coefficient

$$= 185 \times 31.5 \times 16.30 \times 0.827 \times 1.19$$

$$= 93\,481\,\text{m}^3$$

In the Bulk Carrier example, the 1.19 is a hull-form modification coefficient for Bulk Carriers, based on feedback of ships built and service. It is linked with the amount of parallel body designed into this type of vessel.

Questions

1 Define the following cargo capacity terms:
   (a) Moulded Capacity, (b) Grain Capacity, (c) Bale Capacity and (d) Insulated volume.
2 With relationship to grain cargo capacity calculations, list the items that are included in the values 'D_B' or 'D_D.'
3 If the C_B@SLWL is 0.692 for a General Cargo ship, estimate the approximate C_B at 85% of the Depth Mld.
4 Discuss what each term means in the formula for the volume for a Bulk Carrier where volume $= L_h \times B \times D_h \times C_B \times$ hull-form coefficient.
5 The particulars for a General Cargo basic ship and a new design are as follows:

Item	Basic ship	New design
LBP (m)	133	137
Br. Mld (m)	18.36	19.50
Depth Mld (m)	11.55	12.20
SLWL (m)	8.95	9.52
C_B@SLWL	0.745	0.753
Length of amidships	19.50	19.93
Machinery Space (m)		
Tank Top height (m)	1.25	1.42
Upper Deck camber (m)	0.34	0.38
Deck sheer for'd (m)	2.75	2.76
Deck sheer aft (m)	1.45	1.38
Tank Top ceiling (m)	0.065	0.065
Grain Capacity (m³)	17850	xxxx

Estimate the Final Grain Capacity and Bale Capacity for the new design.

Chapter 4

Approximate hydrostatic particulars

When the Naval Architect decides on the final general particulars (as shown in a previous chapter), he can then estimate the hydrostatic values for the new ship.

These values will be for several waterlines, ranging from the fully loaded waterline (Summer Load Waterline, SLWL) down to the lightweight waterline.

The hydrostatic values calculated are as follows:

Displacement	C_B	C_W
KB	BM_T	KM_T
WPA	TPC	MCTC
BM_L	KM_L	GM_T
GM_L	KG	C_B/C_W

Waterplane area (WPA); Tonnes per centimetre immersion (TPC); Moment to change trim one centimetre (MCTC).

Every value except those relating to the (vertical centre of gravity) KG, is dependent upon the Main Dimensions and the *geometrical form* of the new vessel. KG is dependent on how the *loading* of the vessel.

C_B values

The first value required is the C_B. Prior to 1965, R. Munro-Smith suggested *several* formulae for C_B, each one being dependent and changeable with each draft below SLWL.

In 1985, *one* formula was developed that would give the C_B at any draft and for any type of Merchant ship. This formula was developed by the author:

$$\text{Any} \quad C_B = C_B@SLWL \times \left(\frac{\text{Any waterline}}{\text{SLWL}}\right)^x \qquad \text{C.B. Barrass (1985)}$$

where $x = 4.5 \times e^{-5 \times C_B@SLWL}$, $e = 2.718$.

Worked example 4.1

At an 8 m draft SLWL, the $C_B = 0.701$. Calculate the C_B values at drafts of 2.75–9 m:

$$x = 4.5 \times e^{-5 \times C_B @SLWL} = 4.5 \times 2.718^{-5 \times 0.701} = 4.5/33.27 \quad \text{So} \quad x = 0.135$$

$$\text{Any} \quad C_B = C_B @SLWL \times \left(\frac{\text{Any waterline}}{\text{SLWL}} \right)^x$$

$$\text{Any} \quad C_B = 0.701 \times \left(\frac{2.75 \text{ to } 9.0}{8} \right)^{0.135} \text{m}$$

Substituting in each waterline value gives the C_B figures as shown in Table 4.1.

Table 4.1 C_B, C_W and C_B/C_W values at selected waterlines for a new ship

Selected waterline (m)	C_B figures	C_W figures	C_B/C_W values
9.00	0.712	0.812	0.877
8.00 (SLWL)	0.701	0.801	0.875
7.00	0.688	0.788	0.873
6.00	0.674	0.774	0.871
5.00	0.658	0.758	0.868
4.00	0.638	0.738	0.864
3.00	0.614	0.714	0.860
2.75	0.607	0.707	0.858

C_W values

Having now obtained the C_B values, the next step is to evaluate the C_W values at each waterline.

For Merchant ships, *at the SLWL only*:

$$C_W = C_B + K \quad \text{where} \quad K = \frac{1 - C_B}{3} \quad \text{See Figure 4.1}$$

Each ship type will have a particular value for C_B at SLWL and so each ship type will have a different K value. From the SLWL down to the light-weight draft, the C_B and C_W curves will be parallel and be separated by this value K.

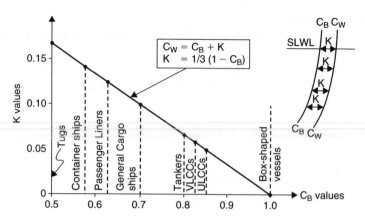

Fig. 4.1 $K \propto C_B$, for SLWL only!!

Worked example 4.2

Calculate the C_W and C_B/C_W values at each waterline for the ship in Worked example 4.1.

$$C_W = C_B + K \quad \text{where} \quad K = \frac{1 - C_B}{3}$$

$$K = \frac{1 - 0.701}{3} = 0.100$$

Another good approximation for the C_W value is that:

$$C_W = \left(\frac{2}{3} \times C_B\right) + \frac{1}{3} \quad \text{but at the SLWL only!!} \quad \text{See Figure 4.2}$$

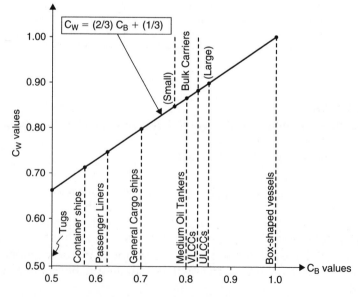

Fig. 4.2 $C_W \propto C_B$, for SLWL only!!

Displacements

For the General Cargo ship under consideration, the displacement (W) is:

$$W = L \times B \times H \times C_B \times p \text{ tonnes, at each waterline of } H$$

$$W = 135.5 \times 18.3 \times H \times C_B \times 1.025 = 2542\,HC_B \text{ tonnes}$$

where H and C_B are variables, changing at each waterline or draft of the new ship.

KB or vertical centre of buoyancy values

This is the vertical centre of buoyancy (VCB) above the base (see Figure 4.3).

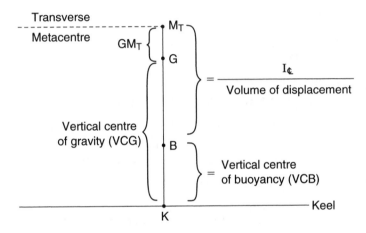

Fig. 4.3   Transverse ship stability factors.

There are various approximations for KB:

1. $KB = H/(1 + C_B/C_w)$,
2. $KB = 0.535 \times H$,
3. $KB = H/2$ for box-shaped vessels,
4. $KB = H \times 2/3$ for triangular-shaped vessels,
5. $KB = 0.700 \times H$ for yachts.

For the vessel being considered in Worked example 4.1, the formula used was:

$$KB = \frac{H}{1 + C_B/C_W} \text{ m, with variables of } H \text{ and } C_B/C_W \text{ at each draft.}$$

BM_T values

For any vessel, $BM_T = \dfrac{I_T}{V}\,m$ See Figure 4.3

where:

I_T = transverse moment of inertia about the waterplane's centreline in m^4,
V = volume of displacement in cubic metres.

For a box-shaped vessel, $BM_T = \dfrac{L \times (B^3/12)}{L \times B \times H \times C_B}$

So $BM_T = \dfrac{B^2}{12 \times H}\,m$

For shipshape vessels, instead of using 1/12, the coefficient 'η' is used. This is an inertia coefficient. 1/12 changes to 'η' because the shape of each waterline is not rectangular.

Hence $BM_T = \dfrac{\eta \times B^2}{H \times C_B}\,m$

where η, H and C_B are variables changing at each waterline.

For Worked example 4.1, $B^2 = 18.3 \times 18.3 = 334.9$

Research by the author has shown that for C_W of 0.692–0.893, for Merchant ships:

$$\eta = 0.084 \times (C_W)^2 \quad \text{C.B. Barrass (1991)}$$

Thus for the vessel being considered:

Any $BM_T = \dfrac{28.13 \times (C_W)^2}{H \times C_B}\,m$,   at each waterline or draft

For example, at 8 m draft, where $C_W = 0.801$, $C_B = 0.701$, then:

$$BM_T = \dfrac{28.13 \times 0.801 \times 0.801}{8.0 \times 0.701}$$

$$BM_T = 3.22\,m$$

'η', the inertia coefficient may also be assumed to be 1/12 or 0.8333, thereby with respect to transverse stability erring slightly on the lower and safer side.

Transverse Metacentre (KM_T)

There are two formulae for KM_T:

$$KM_T = KG + GM_T \quad \text{and} \quad KM_T = KB + BM_T \quad \text{See Figure 4.3}$$

For a ship to be in stable equilibrium, G must be below M_T.

Waterplane area (WPA)

WPA = L × B × C_W m², at each waterline   L = 135.5 m   B = 18.3 m

WPA = 135.5 × 18.3 × C_W

Any   WPA = 2480 × C_W m², at each waterline C_W is the variable.

Tonnes per centimetre immersion (TPC)

$$TPC = \frac{WPA}{100} \times p$$

$$= \frac{WPA}{100} \times 1.025 \quad p = \text{water density of } 1.025 \text{ tonnes/m}^3$$

$$TPC = \frac{WPA}{97.56} \quad \text{tonnes in salt water at each waterline}$$

Moment to change trim one centimetre (MCTC)

For Oil Tankers, MCTC in salt water $= \dfrac{7.8 \times TPC^2}{B}$ tm/cm

For General Cargo ships, MCTC in salt water $= \dfrac{7.31 \times TPC^2}{B}$ tm/cm

For the considered vessel, B = 18.3 m, therefore,

MCTC in salt water $= \dfrac{7.31 \times TPC^2}{18.3}$

$$= 0.399 \times TPC^2 \text{ tm/cm, at each draft}$$

BM_L values

For any vessel, $BM_L = \dfrac{I_{LCF}}{V}$ See Figure 4.4

where:

I_{LCF} = longitudinal moment of inertia of the waterplane, about its longitudinal centre of flotation in m⁴.

V = volume of displacement in cubic metres.

For a box-shaped vessel, $BM_L = \dfrac{B \times (L^3/12)}{L \times B \times H \times C_B}$

Thus $BM_L = \dfrac{L^2}{12 \times H}$ m

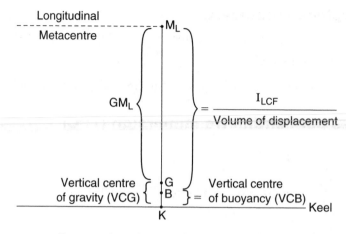

Fig. 4.4   Longitudinal ship stability factors.

For shipshape vessels, instead of using 1/12, the coefficient 'η' is used. This is an inertia coefficient. 1/12 changes to 'η' because the shape of each water-line is not rectangular.

Hence $BM_L = \dfrac{\eta \times L^2}{H \times C_B}$ m

where η, H and C_B are variables changing at each waterline.

For Worked example 4.1, $L^2 = 135.5 \times 135.5 = 18\,360$

Research by the author has shown that for C_W of 0.692–0.893, for Merchant ships:

$$\eta = \frac{3}{40} \times (C_W)^2 \quad \text{C.B. Barrass (1991)}$$

Any $BM_L = \dfrac{1377 \times (C_W)^2}{H \times C_B}$ m

Therefore, for the vessel being considered:
   For the considered example, at 8 m draft, where $C_W = 0.801$, $C_B = 0.701$, then:

$$BM_L = \frac{1377 \times 0.801 \times 0.801}{8.0 \times 0.701}$$

$$BM_L = 157.5 \text{ m}$$

Longitudinal Metacentre values (KM$_L$)

There are two formulae for KM$_L$:

$$KM_L = KG + GM_L \quad \text{and} \quad KM_L = KB + BM_L \quad \text{See Figure 4.4.}$$

For a ship to be in stable equilibrium, G must be below M$_L$.

Hydrostatic curves

Having worked through all of the approximate hydrostatics for the ship's hull form, it is now possible to calculate their values at all drafts, e.g. from 9.0 m down to 2.75 m.

Table 4.3 shows the values obtained at each draft. Figure 4.5 shows the hydrostatic curves obtained after plotting the drafts against the tabulated values.

A set of these tabulated values or a set of hydrostatic curves are supplied to each completed ship.

Nowadays they will probably form part of a computer package of text plus graphics.

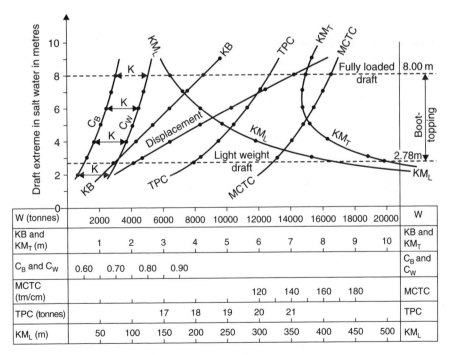

Fig. 4.5   Hydrostatic curves for Worked example (see Table 4.3).

Transverse Metacentric height (GM$_T$)

GM$_T$ is crucial to ship stability. Table 4.2 shows typical safe working values for several ship types at their fully loaded drafts.

Table 4.2   Typical GM$_T$ values for fully loaded conditions at SLWL

Ship type	GM$_T$ at fully loaded draft (SLWL)
General Cargo ships	0.30–0.50 m
Oil Tankers to VLCCs	0.30–1.00 m
Container ships	1.50 m approximately
RO-RO vessels	1.50 m approximately
Bulk Ore carriers	2.00–3.00 m

At drafts below the fully loaded draft, KM$_T$ will usually be larger in value. KG generally will not alter as much. Consequently, GM$_T$ at the lower drafts will be larger in value than those shown in Table 4.2.

For all conditions of loading, the Department of Transport (DfT) stipulate that GM$_T$ must never be less than *0.15 m*. If G is above KM$_T$ then GM$_T$ is said to be negative. If GM$_T$ is negative the ship will be in unstable equilibrium and will capsize.

Some comparisons

Table 4.3 shows the KB value. As previously stated KB could have been calculated using the formula:

Any   KB = 0.535 × any draft   See Table 4.4 for comparable values

Table 4.3 also shows values of draft (H) against WPA values. If $\log_{10}$(WPA) is plotted against $\log_{10}$(H) the result is a straight line graph and the resulting equation resolves to:

$$WPA = 1555 \times H^{0.118} \, m^2 \quad \text{See Table 4.4 for comparable values}$$

In the absence of GM$_L$, the BM$_L$ may be used to estimate the value of MCTC is:

$$MCTC = \frac{W \times GM_L}{100 \times L} \, tm/cm$$

$$= \frac{W \times BM_L}{100 \times L} \, tm/cm \quad L = 135.5 \, m$$

So $MCTC = \dfrac{W \times BM_L}{13\,550} \, tm/cm$ See Table 4.4 for comparable values

Table 4.3 Approximate hydrostatic values for a General Cargo ship, having an LBP of 135.5 m. Also see Figure 4.5 for the resultant hydrostatic curves, plotted from the above values

Draft H (m)	C_B	C_W	W (tonnes)	KB (m)	BM_T (m)	KM_T (m)	BM_L (m)	KM_L (m)	WPA (m²)	TPC (tonnes)	MCTC (tm/cm)	CB/C_W
9.00	0.712	0.812	16 289	4.79	2.89	7.68	141.7	146.5	2014	20.64	170.1	0.877
8.00	0.701	0.801	14 256	4.27	3.22	7.49	157.5	161.8	1986	20.36	165.6	0.875
7.00	0.688	0.788	12 242	3.74	3.63	7.37	177.5	181.2	1954	20.03	160.2	0.873
6.00	0.674	0.774	10 280	3.21	4.17	7.38	204.0	207.2	1920	19.68	154.6	0.871
5.00	0.658	0.758	8363	2.68	4.91	7.59	240.5	243.2	1880	19.27	148.3	0.868
4.00	0.638	0.738	6487	2.15	6.00	8.15	293.9	296.1	1830	18.76	140.5	0.864
3.00	0.614	0.714	4682	1.61	7.79	9.40	381.1	382.7	1771	18.15	131.5	0.860
2.75	0.607	0.707	4243	1.48	8.42	9.90	412.3	413.8	1753	17.97	129.0	0.858

Summary of formulae for Table 4.3:

$C_W = C_B + 0.100$, $K = 0.100$,

$W = 2542 \times H \times C_B$,

$KB = H/(1 + C_B/C_W)$,

$BM_T = (28.13 \times C_W^2)/(H \times C_B)$,

$KM_T = KB + BM_T$,

$BM_L = (1377 \times (C_W)^2)/(H \times C_B)$,

$KM_L = KB + BM_L$,

$WPA = 2480 \times C_W$,

$TPC = 25.4164 \times C_W$,

$MCTC = 258.04 \times (C_W)^2$,

LBP = 135.5 m,

Br. Mld = 18.3 m,

Draft Mld = 8.0 m,

C_B@SLWL = 0.701,

Draft Mld = SLWL.

Table 4.4   Comparison of sets of values for KB, WPA and MCTC

Draft (H) (m)	KB from Table 4.3	KB = $0.535 \times H$	WPA from Table 4.3	WPA = $1555 \times H^{0.118}$	MCTC from Table 4.3	MCTC = $\dfrac{W \times BM_L}{13\,550}$
9.00	4.79	4.82	2014	2015	170.1	170.3
8.00	4.27	4.28	1986	1987	165.6	165.7
7.00	3.74	3.75	1954	1956	160.2	160.4
6.00	3.21	3.21	1920	1921	154.6	154.8
5.00	2.68	2.68	1880	1880	148.3	148.4
4.00	2.15	2.14	1830	1831	140.5	140.7
3.00	1.61	1.61	1771	1770	131.5	131.7
2.75	1.48	1.47	1753	1752	129.0	129.1

Comparison between BM$_T$ and BM$_L$

From previous notes, it can be seen that approximately:

$$BM_T = \frac{0.084 \times (C_W)^2 \times B^2}{H \times C_B} \quad \text{and} \quad BM_L = \frac{3/40 \times (C_W)^2 \times L^2}{H \times C_B}$$

So
$$\frac{BM_T}{BM_L} = 1.12 \times \left(\frac{B}{L}\right)^2$$

Thus $BM_T = BM_L \times 1.12 \times \left(\dfrac{B}{L}\right)^2$ m, at each draft

For the considered ship, L = 135.5 m and B = 18.3 m.

So $BM_T = 0.02043 \times BM_L$ at each draft.

Put another way $BM_L = 0.893 \times BM_T \times \left(\dfrac{L}{B}\right)^2$ m

KG or vertical centre of gravity values

All of the values shown in Table 4.3 and plotted in Figure 4.5 depend on the *geometrical form* of the vessel. They will not alter at each draft unless the vessel undergoes 'ship surgery.' It may be that a new length of midship section is later built and welded into the vessel. More than likely, once the ship has been handed over to the shipowners, these tabulated values will remain un-modified throughout the commercial life of the vessel.

KG however is a different matter. KG depends on the lightweight of the ship together with the condition of loading. It can change from hour to hour or minute by minute, simply by adding, discharging or moving weight vertically about the vessel.

KG will depend upon:

- type of ship;
- type of propelling machinery;
- materials used: metals, woods, plastics, etc.;
- size and layout of accommodation spaces;
- extent of insulated spaces;
- cargo handling arrangements.

It is expedient to determine the KG for the hull weight of the basic ship. This is then proportioned to the Depth Moulded (Mld) for the basic ship. The same proportion is then used for the new design. To this latter figure, insert the values of the new design's machinery weight and its vertical centre of gravity (VCG).

A moment of weight table will then give the final KG value for the new ship's lightweight condition. Each deadweight item can then be introduced each with its respective VCG above the base. The total moment/total weight will give the final KG for that condition of loading.

As a guide, for General Cargo ships, KG of the hull is about 60–70% of the depth Mld.

As a guide, for Shelter Deck vessels, KG is 68–73% of the depth to the uppermost continuous deck.

Relationship between draft, W, C_B and C_W

The 135.5 m ship being considered has a Summer deadweight of 10 000 tonnes. If the fully loaded is as calculated 14 256 tonnes, then the lightweight must be 4256 tonnes. The question then arises, 'what is the draft for this lightweight of 4256 tonnes?'

To obtain the answer to this, a formula suggested by R. Munro-Smith can be used. It is:

$$\frac{H_2}{H_1} = \left(\frac{W_2}{W_1}\right)^{C_B/C_W}$$

For the considered ship: $H_2 = 8\,m$, $W_1 = 14\,256$ tonnes, $W_2 = 4256$ tonnes, C_B/C_W is for the draft of 8 m and is 0.875 in value. H_2 will be the required draft at this ship's lightweight of 4256 tonnes:

$$\frac{H_2}{H_1} = \left(\frac{W_2}{W_1}\right)^{C_B/C_W} \quad \text{So} \quad H_2 = H_1 \times \left(\frac{W_2}{W_1}\right)^{C_B/C_W}$$

$$H_2 = 8 \times \left(\frac{4256}{14\,256}\right)^{0.875}$$

$$H_2 = 2.78\,m$$

Hence, at the lightweight (empty ship) of 4256 tonnes, the light draft is 2.78 m.

Longitudinal centre of buoyancy

This is the centroid of the underwater part of the ship. It is usually measured from amidships. Its value is dependent upon the type of ship. For example, for a Very Large Crude Carrier (VLCC) with a C_B of 0.825, when fully loaded the longitudinal centre of buoyancy (LCB) will be about (2.5% × length between perpendiculars (LBP)) *for'd* of amidships. For fast Passenger Liners with a C_B of say 0.600, it will be about (2.5% × LBP) *aft* of amidships. This is due to the more streamlined hull form with hollowed out forward end contours.

Figure 4.6 shows graphically the LCB position relative to amidships for several other ship types. Generally, the higher the required service speed, the smaller will be the C_B value and the further aft will be the LCB position. Note how for General Cargo ships the LCB is located very near to amidships when fully loaded.

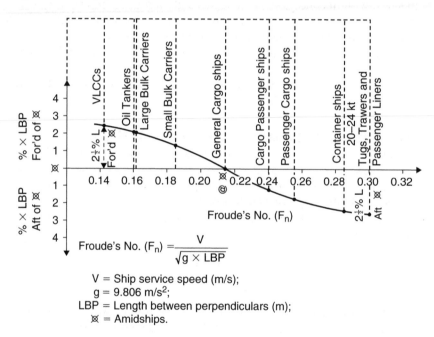

Froude's No. $(F_n) = \dfrac{V}{\sqrt{g \times LBP}}$

V = Ship service speed (m/s);
g = 9.806 m/s²;
LBP = Length between perpendiculars (m);
⊠ = Amidships.

Fig. 4.6   Position of LCB relative to amidships for ships when fully loaded to SLWL.

Summary

At the design stage, it is essential that the Naval Architect gets the Main Dimensions and the design coefficients correct. If he does not, then the hydrostatic values covered in this chapter could give a vessel that possesses unstable equilibrium.

At this stage of the design, a slight increase in the provisional Breadth Mld (Br. Mld) may solve a problem of instability. If the vessel is nearing completion at a shipyard, then this problem can be resolved by introducing permanent solid iron ore ballast in the double-bottom tanks. Such a procedure would lower G in the vessel to being below the Transverse Metacentre M_T.

Mistakes have been made at the design stage. Such vessels have been built and today operate commercially. It can however give the Naval Architect a less than respected reputation. Finally it should be mentioned that some vessels do carry solid ballast in their double bottom tanks. This is because of breadth restrictions, perhaps whilst transiting a canal or lock along their intended route.

Questions

1 For Merchant ships, $C_W = C_B + K$. Suggest approximate K values for:
   (a) Container vessels, (b) Passenger Liners, (c) General Cargo ships,
   (d) Large Bulk Carriers, (e) ULCCs.
2 A VLCC has a Draft Mld of 16.76 m and a Depth Mld of 25.60 m. At 16.76 m draft the C_B is 0.827. Estimate by *two* methods, her C_B at 85% Depth Mld.
3 (a) A Bulk Carrier has an LBP of 182 m, a Br. Mld of 32.26 m, a Draft Mld of 10.75 m with a C_B of 0.787. Calculate the C_B values corresponding to drafts of 10, 8, 6, and 4 m (lightweight draft).
   (b) Calculate the fully loaded displacement, the lightweight and the deadweight in tonnes.
4 $KM_T = KB + BM_T$. Also $KM_T = KG + GM_T$. Explain in detail, what each of these stability factors indicate.
5 A new General Cargo ship has an LBP of 133 m, a Br. Mld of 18 m, a Draft Mld of 7.5 m and a C_B of 0.720 at the 7.5 m draft. For the SLWL, calculate the following hydrostatic data: displacement, C_W, C_B/C_W, KB, WPA, TPC, MCTC.
6 For the vessel in Q5; calculate the C_W, transverse 'η', BM_T, longitudinal 'η', BM_L.

Chapter 5

Types of ship resistance

The resistance of ships can be predicted by several methods. One method that has stood the test of time is by William Froude in 1870 et seq. He used planks and ship models.

Nowadays ship models may be made in wood, paraffin wax or polyurethane foam. Ship models today range from being 3 to 10 m in length.

Resistance can be divided into four groups:

1. Frictional resistance (R_f).
2. Wave-making resistance.
3. Eddy-making resistance.
4. Resistance due to wind and appendages (added on later for full size ships in sea conditions).

When added together, resistances 2 and 3 form the residual resistance (R_r):

Total resistance (R_T) = R_f + R_r for ship models in calm water

Frictional resistance

One of the first men to successfully research into frictional resistance of ships was William Froude. After much research on planks and ship models he suggested a formula for R_f. It was, and is still used today:

$$R_f = f \times A \times V^n N$$

where:

f = a coefficient that depends on length of ship model or vessel, roughness of the hull and density of water in which the vessel is moving,
A = wetted surface area (WSA) of the vessel's hull in square metre,
V = forward speed of the ship model or vessel in knots,
n = a coefficient dependent upon the roughness of the hull.

Having covered each factor in William Froude's formula, it is necessary to proceed to show how each factor may be estimated.

The values of 'f' can be calculated using the following formulae:

For prototypes, $f = 0.441 / L_s^{0.0088}$ for ship lengths of 125–300 m
C.B. Barrass (1991)

For ship models, $f = 0.6234 / L_m^{0.1176}$ for ship model lengths 3–10 m
C.B. Barrass (1991)

where L = length between perpendiculars (LBP) for both ship models and the full size ships, operating in salt water conditions with forward speeds in knots.

Worked example 5.1

Estimate the 'f' coefficient for a 175 m long ship when operating in salt water of density of 1.025 tonnes/m^3:

$$f = 0.441 / L_s^{0.0088} = 0.441 / 175^{0.0088} = 0.421\,41$$

An accurate prediction for the WSA is that suggested by D.W. Taylor of the Washington Test Tank:

$$WSA = A = 2.56 \times (W \times L)^{0.5}\,m^2 \quad \text{D.W. Taylor}$$

where:

W = vessel's displacement in tonnes,
L = vessel's LBP in metres,
WSA is for the ship model or ship in salt water conditions.

Worked example 5.2

Estimate the WSA for a fully loaded General Cargo ship having a displacement of 20 000 tonnes and an LBP of 135 m. Assume salt water conditions:

$$A = 2.56 \times (W \times L)^{0.5}\,m^2 = 2.56 \times (20\,000 \times 135)^{0.5}$$

Therefore A = WSA = 4207 m^2 (the area of the hull in contact with the water in which the ship floats).

For a ship model being tested in a towing tank or in a flume, the forward speed will be of the order of 3 kt. This will correspond geometrically to the forward speed of the full size ship.

For Merchant ships, depending on ship type, the designed service speed (V) will be in the range from 13 to 30 kt.

The roughness coefficient 'n' for modern welded ships (and ship models) has been agreed by 1TTC conference (1957) to be a value of 1.825.

Summarising, for the 175 m ship, at a speed of 16 kt:

$$R_f = 0.42141 \times \{2.56 \times (W_S \times 175)^{0.5}\} \times 16^{1.825}\,N \quad \text{in salt water}$$

Note: If W is changed from tonnes to cubic metre the coefficient changes from being 2.56–2.59. Consequently, the fourth term becomes the square root of m^4. This gives units of square metre, which are the required units for area waterplane area (WPA).

Froude's speed–length law

In his research work into ship resistance, William Froude had to decide upon a relationship between the full size vessels he was considering and the corresponding ship models he was testing. After much investigation he decided to use:

$$V_s/L_s^{0.5} = V_m/L_m^{0.5} \quad \text{known as Froude's speed–length law}$$

where:

suffix 's' is for ship, suffix 'm' is for model of ship,
V = velocity or speed of ship or ship model in knots,
L = LBP of ship or ship model in metres.

Worked example 5.3

A ship model is 4.84 m LBP. It is towed at a speed of 3 kt. The full size ship is to be 121 m LBP. Estimate the corresponding ship's speed in knots:

$$V_s/L_s^{0.5} = V_m/L_m^{0.5} \quad \text{So} \quad V_s/121^{0.5} = 3/4.84^{0.5}$$

$$V_s = 3 \times 25^{0.5} = 15\,\text{kt}$$

Worked example 5.4

Speed of a ship model is 3.3 kt. Ship's LBP is 190 m. Designed service speed of ship is to be 18 kt. Using Froude's speed–length law, estimate the corresponding LBP of the ship model:

$$18/190^{0.5} = 3.2/L_m^{0.5} \quad \text{Thus} \quad L_m = 3.2 \times 190^{0.5}/18$$

So L_m = length of the ship model = 6 m

Froude number

As well as a Froude speed–length law, there is a Froude Number (Fn). This is dimensionless and is given by:

$$Fn = \frac{V}{(g \times L)^{0.5}} = \frac{0.319 \times V}{L^{0.5}}$$

where:

V is vessel speed now in m/s,
L is in metres,
g = gravity = 9.806 m/s^2.

Relationship between Froude's speed–length law and Froude's Number

$$V/L^{0.5} = 6.0871 \times Fn \quad \text{So} \quad Fn = 0.16428 \times V/L^{0.5}$$

The units are as previously stated.

Residual resistance

From his speed–length law, William Froude was able to develop relationships between the residual resistances of his ship models and the corresponding full size ships. By using his speed–length law, he was able to demonstrate that:

$$\frac{R_r(\text{ship})}{R_r(\text{model})} = \left(\frac{L_s}{L_m}\right)^3$$

From Applied Mechanics, it can be shown that for geometrically similar shapes (geosims), that volumes vary as their lengths cubed. Combining these facts together it follows that:

$$\frac{R_r(\text{ship})}{R_r(\text{model})} = \left(\frac{L_s}{L_m}\right)^3 = \frac{\text{Volume (ship)}}{\text{Volume (model)}}$$

Hece for geosims, $R_r' \propto L^3$ and that $R_r \propto$ Volume of displacement.

Worked example 5.5

A ship model is 7 m LBP and is towed at 3.25 kt. The prototype is 140 m LBP. Estimate the full size ship's residual resistance if the model's residual resistance is 16 N. Also calculate the corresponding ship speed:

$$\frac{R_r(\text{ship})}{R_r(\text{model})} = \left(\frac{L_s}{L_m}\right)^3 \quad \text{So} \quad R_s(\text{ship}) = 16 \times \left(\frac{140}{7}\right)^3$$

$$= 128000\,\text{N} = 128\,\text{kN}$$

$$V_s/L_s^{0.5} = V_m/L_m^{0.5} \quad \text{So} \quad V_s = 3.25 \times \left(\frac{140}{7}\right)^{0.5}$$

$$V_s = \text{ship's speed} = 14.6\,\text{kt}$$

Total resistance

$R_T = R_f + R_r$ in calm waters, with no wind and appendage allowances

For RO-RO ships and Passenger Liners $R_f/R_T = 30\%$ approximately

For Container vessels $R_f/R_T = 40{-}50\%$ approximately

For General Cargo ships $R_f/R_T = 60{-}70\%$ approximately

For Supertankers $R_f/R_T = 90\%$ approximately

Figure 5.1 graphically shows percentages for other Merchant ships. Note how vessels with higher service speeds have lower values for R_f/R_T. This is because extra wave-making resistance occurs. Consequently the frictional part of the increased total resistance is by comparison of a lower percentage. This argument is shown to be particularly true for the larger slower moving Supertankers.

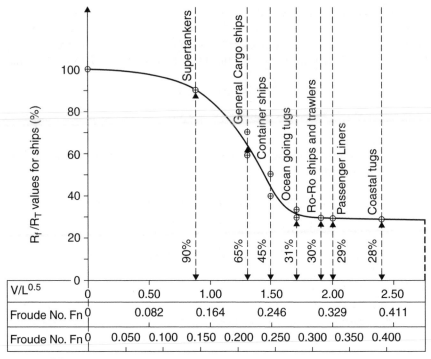

For $V/L^{0.5}$; V in knots, L in metres
For   Fn of $V/(gL)^{0.5}$; V (m/s), g = 9.806 m/s², L = LBP in metres
V = ship's speed

Fig. 5.1 R_f/R_T for several Merchant ships when at loaded departure's draft moulded.

Wind and appendage allowances

This will very much depend upon the intended route for the new vessel. It is usual practice for Naval Architects to add 10–30% onto the R_T obtained for calm water conditions. The higher percentages will appertain for vessels operating on the more heavy weather routes.

Worked example 5.6

A 7.32 m ship model has a WSA of 6.31 m². It is towed in fresh water at a speed of 3 kt. The total resistance is measured on the model and found to be 32 N.

(a) Calculate R_T for a ship of 144 m LBP in calm water conditions. Assume 'n' = 1.825 for the ship model and prototype.
(b) If the wind and appendage allowances total 22%, then proceed to estimate the final total resistance in kN in sea conditions.

First modify model's R_T for salt water density:

$$R_T \text{ for model} = 32 \times (1.025/1.000) = 32.80 \text{ N}$$

$$R_f \text{ for model} = f \times A \times V^n$$

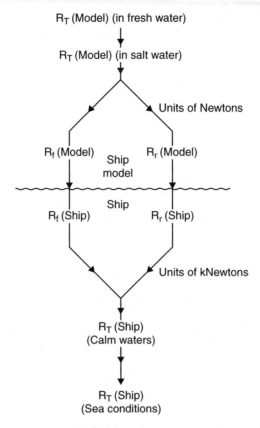

Fig. 5.2   Resistance line diagram for Worked example 5.6.

$$= (0.6234/7.32^{0.1176}) \times 6.31 \times 3^{1.825}$$
$$= 23.11\,N$$

For the model, $R_T = R_f + R_r$ So $R_r = R_T - R_f$

Thus for the model, $R_r = 32.80 - 23.11$

$$R_r = 9.68\,N \text{ in salt water}$$

$Rr(ship) / Rr(model) = (Ls / LM)^3$ So $Rr(ship) = Rr(model) \times (Ls / Lm)^3$
$Rr(ship) = 9.69 \times (144 / 7.32)^3 = 73{,}770\,N = 73.77\,kN$ in salt water.

To find speed of the ship

$$V_s / L_s^{0.5} = V_m / L_m^{0.5} \quad \text{So} \quad V_s = 3 \times \left(\frac{144}{7.32}\right)^{0.5}$$

Thus $V_s = 13.28\,kt$ = service speed of the ship.

To find the wetted service area of the ship

For geosims, it is known that volumes vary as the lengths cubed and areas vary as the lengths squared:

Hence $\quad \dfrac{\text{WSA(ship)}}{\text{WSA(model)}} = \left(\dfrac{L_s}{L_m}\right)^2$

Thus $\quad \text{WSA(ship)} = 6.31 \times \left(\dfrac{144}{7.32}\right)^2 = 2440\,\text{m}^2$

For the ship, $R_f = f \times A \times V^n$

Thus $\qquad R_f = (0.441/144^{0.0088}) \times 2440 \times 13.28^{1.825}$

$\qquad\qquad = 0.4221 \times 2440 \times 112.16$

$\quad R_f(\text{ship}) = 115\,516\,\text{N} = 115.52\,\text{kN}$

In calm waters, $R_T = R_f + R_r = 115.42 + 73.77$

Thus for the ship, $R_T = 189.22\,\text{kN}$ in salt water

Note in passing that $R_f/R_T = 115.52/189.22 = 0.6105 = 61.05\%$. This compares favourably with the General Cargo ships in Figure 5.1.

In weather conditions at sea, $R_T = 189.22 \times 1.22/1.00$. This is because the wind and appendage allowances were given as being $+22\%$:

$$R_T \text{ for the ship} = 189.22 \times 1.22 = 230.8\,\text{kN}.$$

This is the final total resistance for the ship. It is based on towed ship model results in a ship model tank extrapolated to a finished full size ship operating in sea conditions.

Naked effective power

This is the tow rope power of a ship in calm water conditions, without any weather or appendage allowances. It is measured in kW:

Naked effective power $P_{NE} = R_T \text{ (in kN)} \times V \text{ (in m/s)} \qquad 1\,\text{kt} = 1852\,\text{m/s}$

So $\qquad\qquad P_{NE} = 189.22 \times \left(13.28 \times \dfrac{1852}{3600}\right)$

Therefore $\qquad P_{NE} = 1293\,\text{kW}$

Procedure steps for solving ship resistance and P_{NE} problems

1. Draw a line diagram (see Figure 5.2).
2. Modify R_T of the model for density of salt water.
3. Calculate R_f and then R_r for the ship model in salt water.

4. Calculate R_r for the ship using the lengths cubed format.
5. Estimate ship speed using Froude's $V/L^{0.5}$ law.
6. Calculate WSA using the lengths squared format.
7. Using the obtained WSA and ship speed values, estimate R_f for the ship.
8. Add R_f to R_r to obtain R_T for the ship, in calm water conditions.
9. Use the given wind and appendage allowances to obtain R_T for the ship in sea conditions.
10. Finally calculate the P_{NE} in kW.

Three more important geosim relationships

1. For ship models and ships:

$$\frac{\text{Velocity (ship)}}{\text{Velocity (model)}} = \left(\frac{L_s}{L_m}\right)^{0.5} = \left\{\frac{\text{Volume (ship)}}{\text{Volume (model)}}\right\}^{1/2 \times 1/3}$$

Thus
$$\frac{\text{Velocity (ship)}}{\text{Velocity (model)}} = \left\{\frac{\text{Volume (ship)}}{\text{Volume (model)}}\right\}^{1/6}$$

Hence   velocities $\propto$ (volume of displacement)$^{1/6}$ for ship models and ships

2. For geosim ship models:

$$R_f = f \times A \times V^n = \left(\frac{0.6234}{L_m^{0.1176}}\right) \times L^2 \times L^{1.825/2}$$

Thus $R_f = L^{2.000 + 0.9125 - 0.1176}$

Hence $R_f \propto L^{2.7949}$ for geosim ship models

3. For geosim full size ships:

$$R_f = f \times A \times V^n = \left(\frac{0.441}{L_s^{0.0088}}\right) \times L^2 \times L^{1.825/2}$$

Thus $R_f = L^{2.000 + 0.9125 - 0.0088}$

Hence $R_f \propto L^{2.9037}$ for geosim full size ships

Worked example 5.7

A ship has an LBP of 125 m. She has a frictional resistance of 129.02 kN and a residual resistance of 140.08 kN. Calculate the frictional and residual resistances for a geosim new design having an LBP of 130 m:

For ships, $R_f \propto L^{2.9037}$ Thus $\dfrac{R_{f2}}{R_{f1}} = \left(\dfrac{L_2}{L_1}\right)^{2.9037}$

So $R_{f2} = 129.02 \times \left(\dfrac{130}{125}\right)^{2.9037} = 144.58\,\text{kN}$ for the 130 m ship

For ships, $R_r \propto L^3$ Thus $\dfrac{R_{r2}}{R_{r1}} = \left(\dfrac{L_2}{L_1}\right)^3$

So $R_{r2} = 140.08 \times \left(\dfrac{130}{125}\right)^3 = 157.58\,\text{kN}$ for the 130 m

Worked example 5.8

A 5 m ship model has a frictional resistance of 13.32 N. Calculate the frictional resistance for a geosim model having an LBP of 5.5 m:

For ships models, $R_f \propto L^{2.7949}$ Thus $\dfrac{R_{f2}}{R_{f1}} = \left(\dfrac{L_2}{L_1}\right)^{2.7949}$

So $R_{f2} = 13.32 \times \left(\dfrac{5.5}{5.0}\right)^{2.7949} = 17.39\,\text{kN}$ for the 5.5 m model

Questions

1 A ship's model is 6.5 m long and the prototype is 130 m LBP. Calculate the frictional coefficient 'f' for both vessels.
2 A ship's displacement is 14 020 tonnes with an LBP of 125 m. Estimate the WSA in square metre.
3 Ship model speed is 3 kt with an LBP of 7 m. Estimate the speed of a geosim full size ship having an LBP of 150 m.
4 (a) Ship speed is 20 kt with an LBP of 175 m. Calculate her Fn.
   (b) List the four components of ship resistance.
5 Sketch the line diagram for solving ship resistance problems. Label the important points on the diagram.
6 The frictional resistance of a 7.5 m ship model is estimated to be 25.63 N. Calculate R_f of a geosim ship model having an LBP of 8.25 m.
7 The frictional resistance of a 142.5 m length ship is 112.75 m. Calculate Fn of a geosim ship having an LBP of 127.5 m.
8 A 7.55 m length ship model has a WSA of 6.75 m². It is towed through fresh water at a speed of 3 kt. The total resistance is measured and found to be 34 N. Calculate the corresponding speed and the P_{NE} for a ship having an LBP of 148 m operating in sea water.

Chapter 6

Types of ship speed

When dealing with ships, there are three different speeds to consider. They are V_T, V_S and V_a. The following notes how they are interconnected together with the Apparent Slip and the Real Slip:

- V_T is the theoretical speed produced by the ship's propeller working in an *unyielding* fluid. The water just ahead of the propeller is considered to be stationary.

$$V_T = \text{Propeller pitch (P)} \times \text{Propeller revolutions (N)} \times \frac{60}{1852}$$

$$1\,\text{kt} = 1852\,\text{m/s}$$

Hence $V_T = P \times N \times \dfrac{60}{1852} = \dfrac{P \times N}{30.867}\,\text{kt}$

Pitch (P) is the distance moved forward after one complete revolution of the propeller in an unyielding fluid.

- V_S is the speed of ship working in a *yielding* fluid and is the speed of the ship over the ground. In practice, the propeller is working in a fluid that has a forward speed. Because of this there is an energy loss and so V_S is less than V_T. It should be noted that in a ship's technical specification, the speed quoted is this speed V_S. Consequently, V_S is, in effect, the design service speed for a ship.
- V_a is the velocity of advance and is the speed of the ship relative to the water in which the ship is moving. It includes current effects for and against the forward motion of the ship.

Using Figure 6.1, it can be observed that:

$$\text{Apparent Slip} = (V_T - V_S)\,\text{kt}$$

$$\text{Apparent Slip ratio} = \frac{V_T - V_S}{V_T} \quad \text{usually given as a percentage}$$

Apparent Slip ratio may be positive or negative. It may range from -15% to $+15\%$ and so, depending on direction of the current, may be negative, zero or positive.

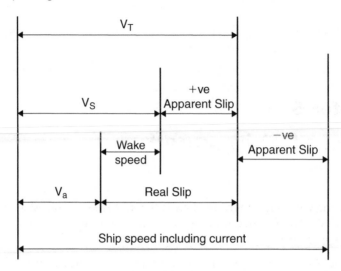

Fig. 6.1   Three ship speeds, Apparent Slip and Real Slip.

$$\text{Real Slip} = (V_T - V_a)\ kt$$

$$\text{Real Slip ratio} = \frac{V_T - V_a}{V_T} \quad \text{usually given as a percentage}$$

Real Slip ratio is always a positive percentage. It is always much greater than the Apparent Slip ratio. In some cases it could be as high as 40%.

$$\text{Wake speed} = (V_S - V_a)\ kt$$

$$\text{Wake speed fraction} = W_t = \frac{V_S - V_a}{V_S} \quad \begin{array}{l}\text{usually given to three decimal} \\ \text{figures}\end{array}$$

A good approximation for W_t is:

$$W_t = (C_B/2) - 0.05 \text{ approximately} \qquad \text{as per D.W. Taylor}$$

where C_B = the ship's block coefficient see Table 6.1.

Table 6.1   Approximate wake fraction values for several Merchant ships

Ship type	Typical C_B, when fully loaded	Approximate W_t
Supertankers	0.825	0.363
Oil Tankers	0.800	0.350
Large Bulk Carriers	0.825	0.363
Small Bulk Carriers	0.775	0.338
General Cargo ships	0.700	0.300
Passenger Liners	0.625	0.263
Container ships	0.575	0.238

Worked example 6.1

A propeller revolves at 120 rpm. It has a pitch of 4.5 m. Ship speed V_S is 15.5 kt. Block coefficient is 0.726.

Calculate the V_T, V_S, V_a, Apparent Slip ratio, Real Slip ratio and wake speed. Make a sketch and where appropriate insert your calculated values.

$$V_T = \frac{P \times N}{30.866} = \frac{4.5 \times 120}{30.867} = 17.49 \text{ kt}$$

$$\text{Apparent Slip} = V_T - V_S = 17.49 - 15.50 = 1.99 \text{ kt}$$

$$\text{Apparent Slip ratio} = \frac{V_T - V_S}{V_T}$$

$$W_t = (C_B/2) - 0.05 \text{ approximately} = (0.726/2) - 0.05 = 0.313$$

$$W_t = \text{Wake speed fraction} = w_t = \frac{V_S - V_a}{V_S} = 0.313$$

$$\text{Thus} \quad 0.313 = \frac{15.50 - V_a}{15.50} \quad \text{So} \quad V_a = 10.65 \text{ kt}$$

$$\text{Check:} \quad \frac{15.50 - 10.65}{15.50} = 0.313 = \text{Wake speed fraction value}$$

$$\text{Real Slip} = V_T - V_a = 17.49 - 10.65 = 6.84 \text{kt}$$

$$\text{Real Slip ratio} = \frac{V_T - V_a}{V_T} = \frac{17.49 - 10.65}{17.49} = +0.3911 = +39.11\%$$

$$\text{Wake speed} = V_S - V_a = 15.50 - 10.65 = 4.85 \text{ kt}$$

Worked example 6.2

A vessel of 12 400 tonnes displacement is 120 m long, 17.5 m beam and floats at an even keel draft of 7.5 m in salt water of density 1.025 tonnes/m³. The propeller has a face pitch ratio of 0.75. When the propeller is turning at 100 rpm, the ship speed (V_S) is 12 kt with a Real Slip ratio of 30%.

Calculate the following: block coefficient, wake fraction, velocity of advance (V_a), theoretical speed (V_T), propeller diameter and the Apparent Slip ratio.

$$\text{Assume that pitch ratio} = \frac{\text{Propeller pitch}}{\text{Propeller diameter}}$$

$$C_B = \frac{\text{Volume of displacement}}{L \times B \times d} = \frac{12\,400/1.025}{120 \times 17.5 \times 7.5}$$

$$\text{Thus} \quad C_B = 0.768$$

$$W_t = \text{wake fraction} = (C_B/2) - 0.05 = (0.768/2) - 0.05 = 0.334$$

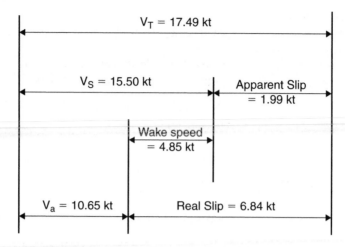

Fig. 6.2   Diagram of the various speeds for Worked example 6.2.

Also
$$W_t = \frac{V_S - V_a}{V_S} = 0.334 \qquad V_S = 12 \, kt$$

$$0.334 = \frac{12 - V_a}{12} \quad So \quad V_a = 7.99 \, kt$$

Check:
$$\frac{12 - 7.99}{12} = 0.334, \quad as \ before$$

Real Slip ratio $= 0.300$ (as given) $= \dfrac{V_T - V_a}{V_T} \qquad thus \quad V_T = 11.41 \, kt$

Check:
$$\frac{11.41 - 7.99}{11.41} = 0.300 = 30\% \ (as \ given)$$

$$V_T = \frac{P \times N}{30.867} = 11.41 \, kt \qquad N = 100 \ (as \ given)$$

So   propeller pitch $= \dfrac{11.41 \times 30.867}{100} = 3.52 \, m$

Pitch ratio $= 0.75$ (as given) $= \dfrac{Pitch}{Diameter}$

So   propeller diameter $= \dfrac{Pitch}{0.75} = \dfrac{3.52}{0.75} = 4.69 \, m$

Apparent Slip ratio $= \dfrac{V_T - V_S}{V_T} = \dfrac{11.41 - 12.00}{11.41}$

$$= -0.0517 = -5.17\%$$

Note how this condition has produced a negative Apparent Slip ratio.

Questions

1 Sketch a diagram to clearly show the theoretical speed (V_T), the ship speed (V_S) and the velocity of advance (V_a). On the diagram, show the Apparent Slip, the Real Slip and the wake speed.

2 A ship's propeller has a pitch of 4.76 m and revolves at 107 rpm. Calculate the theoretical speed V_T in knots.

3 D.W. Taylor suggested a formula for the wake fraction (W_t). Give this formula and derive the approximate W_t values for three different types of Merchant ships.

4 When a propeller of 4.8 m pitch turns at 110 rpm, the Apparent Slip ratio is found to be ($-S\%$) and the Real Slip ratio is found to be ($+1.5 \times S\%$). If the wake speed is 25% of the ship speed (V_S), then calculate the Apparent Slip ratio and the Real Slip ratio.

   Hint: Make a sketch showing the various speeds and then insert your answers as you proceed along with your calculations.

Chapter 7

Types of power in ships

When a ship generates a certain power within the Engine Room, this power will be transmitted along the propeller shaft and eventually to the tips of the propeller blades.

There will be several losses of power enroute as shown in the following treatise. Figure 7.1 shows the powers between the Engine Room and the propeller tips. All powers today are measured in kW:

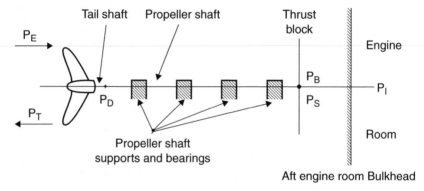

Fig. 7.1   Location of powers along a propeller shaft.

For older ships, any horsepower $\times \frac{3}{4}$ = power in kW approximately.

P_E = Effective power.
P_T = Thrust power = Thrust $\times V_a$.
P_D = Delivered power = $2\pi NT$.
P_S = Shaft power.
P_B = Brake power.
P_I = Input power or Indicated power.
V_a = Velocity of Advance or ship speed relative to the water.
T = Torque.
N = rpm of the propeller.

For this chapter, assume that this theory is for a foreign-going ship, greater than 120 m LBP (length between perpendiculars). As stated in Chapter 5 on ship resistance, the resistance can be predicted from towed ship model tests in towing tanks or flumes.

The measured results will be for calm water conditions. To these results an allowance of 10–30% must be added. This is to account for wind and appendage allowances. The greater percentage relates to vessels intended for the greater heavy weather routes.

The power obtained from the ship model tests is known as the tow-rope power or the naked effective power (P_{NE}) where:

$$P_{NE} = R_T \times V_S \quad \text{in kW}$$

where:

R_T = total ship resistance in calm waters in kW,
V_S = ship speed in m/s.

A more realistic power is the effective power (P_E), where:

$$P_E = P_{NE} + \text{(weather and appendage allowances)}$$

When tested, the towed ship model has a smooth clean hull with no appendages such as bossing, rudder, propeller or bilge keels.

$$P_E = P_{NE} + (10\text{–}30\%) \times P_{NE}$$

Hence $P_E = (1.10\text{–}1.30) \times P_{NE}$ in kW (sea conditions).

Due to the propeller revolving within the sternframe, a vacuum is created. This causes a slight loss in the hull efficiency of the ship. Hull efficiency connects the effective power and the thrust power as follows:

$$P_E / P_T = \text{hull efficiency} \qquad \text{Usually 98–99\%}$$

The proportion or percentage relating the thrust power and the delivered power is the propeller efficiency. This percentage can be as high as 75% for some ships and as low as 60% for vessels such as Supertankers. Its value depends on many variables. See Chapter 9 for a more detailed explanation:

Hence $P_T / P_D = \text{Propeller efficiency} \qquad 60\text{–}75\%$

Efficiency losses occur along the propeller shaft from the tail shaft, forward to the thrust block. These shaft losses will be due to:

- length of the propeller's shaft,
- friction in the shaft support bearings,
- transmission losses due to the propeller shaft material itself.

Hence $P_D / P_B = \text{Shaft efficiency} \qquad \text{for Diesel machinery}$

Hence $P_D / P_S = \text{Shaft efficiency} \qquad \text{for Steam Turbine machinery}$

Obviously, the position of the Engine Room along the ship's length will have a bearing on the final value for this shaft efficiency. Table 7.1 gives an indication of this.

Table 7.1   Propeller shaft efficiency related to the longitudinal position of Engine Room

Position of Engine Room	Approximate shaft losses (%)	Shaft efficiency (%)
Amidships	5.0	95.0
¾ L-Aft	3.5	96.5
All-Aft	2.0	98.0

All-Aft Engine Rooms are fitted on Oil Tankers. ¾ L-Aft Engine Rooms are fitted on many General Cargo ships and Container ships. When located at ¾ L-Aft instead of All-Aft or at amidships, the vibration and longitudinal stresses are reduced.

Note how P_B and P_S are measured at the thrust block. Figure 7.1 shows the thrust block to be outside of the Engine Room. This is not always so. Quite a few ships have the thrust block situated within the Engine Room or Machinery Spaces.

The relationship between the power measured at the thrust block and the input power gives the mechanical efficiency of the ship's engine. It is used to be known as the mechanical advantage where:

$$\text{Mechanical advantage} = \frac{\text{Work in}}{\text{Work out}}$$

Updates of this formula used nowadays are:

$$P_B/P_I = \text{Engine's mechanical efficiency}$$
$$87.5\text{–}92.5\% \text{ for Diesel machinery}$$

$$P_S/P_I = \text{Engine's mechanical efficiency}$$
$$\text{About } 85\% \text{ for Steam Turbine machinery}$$

Worked example 7.1

For a new design, it was found from towed ship model tests that the naked effective power (P_{NE}) for the prototype was 3200 kW. Using a basic ship, it was decided to use the following data for this new design:

Hull efficiency = 99.2%, propeller efficiency = 70.85%, shaft losses = 4.75%. Engine efficiency = 86.13%, weather and appendage allowances = +10%. Steam Turbine machinery installed with thrust block fitted Aft of Engine Room.

Calculate all the powers from the propeller tips to the Engine Room.

First draw a diagram and place the various powers as shown in Figure 7.2. As each power is estimated, insert it onto this diagram. Figure 7.3 shows the completed diagram with all powers in place.

$$P_E = P_{NE} + \text{(weather and appendage allowances)}$$
$$P_E = 3200 + (10\% \times 3200) = 1.10 \times 3200 = 3520\,kW$$

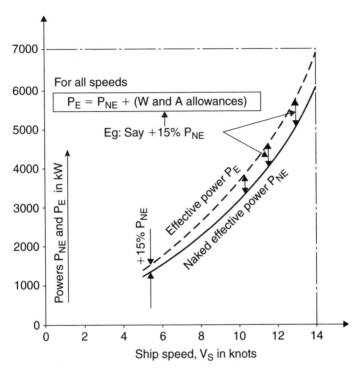

Fig. 7.2   Powers (P_{NE} and P_E) against ship speed (W and A denotes weather and appendage allowances).

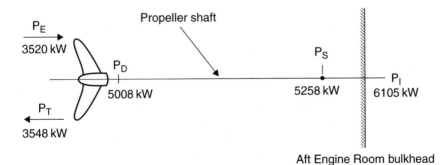

Fig. 7.3   Powers for Worked example 7.1.

$$P_E/P_T = \text{Hull efficiency} = 99.2\% \quad \text{Thus} \quad P_T = P_E/0.992$$
$$= 3520/0.992 = 3548\,\text{kW}$$

$$P_T/P_D = \text{Propeller efficiency} = 70.85\%$$

Thus
$$P_D = P_T/0.7085$$
$$= 3548/0.7085$$
$$= 5008\,\text{kW} \quad \text{at the propeller tail shaft}$$

$$P_D/P_S = \text{Propeller shaft efficiency} = 100\% - 4.75\% = 95.25\%$$

Hence
$$P_S = P_D/0.9525$$
$$= 5008/0.9525$$
$$= 5258\,\text{kW} \qquad \text{at the thrust block}$$

$$P_S/P_I = \text{Engine efficiency} = 86.13\%$$
$$P_I = P_S/0.8613$$
$$= 5258/0.8613$$
$$= 6105\,\text{kW} \qquad \text{in the Engine Room}$$

To check this final answer for the input power, the following formula can be used:

$$P_I = \frac{P_{NE} + (\text{weather and appendage allowances})}{Y}$$

where
$Y = \text{hull efficiency} \times \text{propeller efficiency} \times \text{shaft efficiency} \times \text{engine efficiency}.$

Hence $\quad P_I = \dfrac{3520}{0.992 \times 0.7085 \times 0.9525 \times 0.8613}$

Thus $\quad P_I = 6105\,\text{kW}$ in the Engine Room, as previously evaluated.

Important points to observe in the Worked example 7.1:

1. The input power within the Engine Room needed to be almost *twice* the power extrapolated from ship model tests, i.e. 6105 kW, compared to 3200 kW.
2. Between the input power and the thrust block power, losses amounted to 847 kW.
3. Losses along the propeller shaft resulted in 250 kW.
4. Power losses due to the propeller's efficiency of 70.85% were 1460 kW.
5. Hull losses were small, only amounting to 28 kW.
6. Weather and appendage losses were given as only 10% of the P_{NE}. However if this vessel had been trading on really heavy weather routes, the requirement for P_I within the Engine Room could have been as high as 7215 kW instead of 6105 kW.

Questions

1 Sketch the propeller shaft, from the propeller itself to the Engine Room. On the sketch, label the positions of the ship powers P_E, P_T, P_D, P_B, P_S and P_I.

2 What are the formulae for the thrust power and the delivered power? What value is obtained when the thrust power is divided by the delivered power?

3 Give typical values of the propeller shaft efficiency for Merchant ships. What mostly influences the value of the propeller efficiency?

4 Discuss how the naked effective power is related to the effective power. Include percentage values for weather and appendage allowances.

5 For a new design, it was found that after towing a ship model that the power extrapolated to the full size ship was 3475 kW. Using basic ship information it was decided to use the following information: hull efficiency = 99.24%, propeller efficiency = 68.75%, shaft losses = 2.85%, diesel engine efficiency = 88.73%, weather and appendage allowances = 18.5%.

   (a) Calculate all the powers from the propeller tips to the Engine Room.

   (b) What is the power loss in kW between the thrust block to the propeller tail shaft?

Chapter 8

Power coefficients on ships

When considering powers for ships, Naval Architects use power coefficients. This is to assist them when making quick comparisons between basic ships and new designs. The three coefficients discussed in this chapter are:

1. Quasi-Propulsive Coefficient (QPC).
2. Propulsive Coefficient (PC).
3. Admiralty Coefficient (A_C).

To begin, first please refer to Figure 8.1. This shows the powers along a propeller shaft.

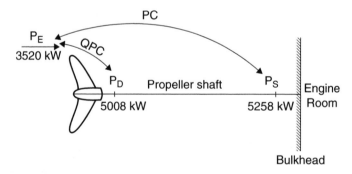

Fig. 8.1   Powers and power coefficients along a propeller shaft.

Quasi Propulsive Coefficient
QPC relates the effective power at the propeller tips with the delivered power located at the tail shaft. It will be close in value to that obtained for the propeller efficiency:

$$QPC = P_E/P_D \qquad about \quad 0.600–0.700 \qquad See\ Figure\ 8.1$$

Dr A. Emmerson of Newcastle upon Tyne University suggested an approximate formula for QPC. This is as follows:

$$QPC = 0.850 - \frac{N \times L^{0.5}}{10\,000} \quad \text{approximately} \quad \text{A. Emmerson}$$

where:

N = propeller rpm,
L = length between perpendiculars (LBP) in metres.

Worked example 8.1
Estimate the QPC for a new design if the LBP is 144 m and the propeller revolutions are 120 rpm. Calculate the answer to three decimal figures.

$$QPC = 0.850 - \frac{120 \times 144^{0.5}}{10\,000} = 0.850 - 0.144$$

Hence $QPC = 0.706$

Propulsive Coefficient
PC relates the effective power with the power measured at the thrust block. It will depend upon the hull efficiency, the propeller efficiency and the propeller shaft efficiency:

$$PC = \frac{P_E}{\text{Power at the thrust block}} \quad \text{see Figure 8.1}$$

$$PC = \frac{P_E}{P_B} \quad \text{for Diesel Machinery}$$

$$PC = \frac{P_E}{P_S} \quad \text{for Steam Turbine installation}$$

Values for PC will be of the order of 0.600–0.700 on Merchant ships.

Worked example 8.2
Calculate the QPC and the PC if the effective power is 3520 kW, the Delivered power is 5008 kW and the shaft power is 5258 kW.

$$QPC = P_E/P_D = 3520/5008 = 0.703$$

$$PC = P_E/P_S = 3520/5258 = 0.669$$

Admiralty Coefficient
If two ships are similar in type, displacement, power and speed, then their A_C values will be similar in value:

$$A_C = \frac{W^{2/3} \times V^3}{P} \quad \begin{array}{l}\text{300–600: with the higher values being for the more} \\ \text{efficient vessels}\end{array}$$

where:

W = ship's displacement in tonnes,
V = ship's speed in knots, for best comparisons V should be <20 kt,
P = power measured at the thrust block,
= P_B for Diesel machinery,
= P_S for Steam Turbine installation.

Worked example 8.3

For a basic ship, the displacement is 14 500 tonnes, service speed is 16 kt and the brake power is 5000 kW. Estimate the A_C.

$$A_C = \frac{W^{2/3} \times V^3}{P_B} = \frac{14\,500^{2/3} \times 16^3}{5000}$$

Hence $A_C = 487$

Worked example 8.4

Vessel	Displacement (tonnes)	Brake power (kW)	Service speed (kt)
Basic ship	14 500	5000	16.00
New design	14 750	5548	V_D

Estimate the service speed (V_D) for the new design.

$A_C(1) = A_C(2)$

Hence $\dfrac{W^{2/3} \times V^3}{P_B}$ for basic ship $= \dfrac{W^{2/3} \times V^3}{P_B}$ for new design

So $\dfrac{14\,500^{2/3} \times 16^3}{5000} = \dfrac{14\,750^{2/3} \times V_D^3}{5548}$

Therefore $4492 = (V_D)^3$ Thus $V_D = 4492^{1/3}$

So $V_D = 16.50\,\text{kt}$ i.e. the service speed for the new design.

An approximation for Admiralty Coefficient

Another way of obtaining the A_C value is to use the approximation of Dr A. Emmerson of Newcastle upon Tyne University. He suggested:

$$A_C = 26\left(L^{0.5} + 150/V\right) \text{ approximately} \text{ A. Emmerson}$$

As before, L = LBP in metres, V = ship speed in knots.

Worked example 8.5

For an SD14 Cargo ship, LBP is 148 m and the service speed is 15 kt. What is the approximate A_C value?

$$A_C = 26\left(L^{0.5} + \frac{150}{V}\right) \text{ approximately} = 26\left(148^{0.5} + \frac{150}{15}\right)$$

$$A_C = 26(12.166 + 10) = 576$$

A note of caution … if the service speed is *20 kt or greater*, it is more accurate when making comparisons, to change the velocity indice from being *three* to being *four*.

Consequently:

$$A_C = \frac{W^{2/3} \times V^4}{P} \qquad \text{for fast ships}$$

Worked example 8.6

A Container ship has a displacement of 16 000 tonnes, a speed of 22 kt and a shaft power of 13 500 kW. Using the formula for fast ships, calculate the A_C value.

$$A_C = \frac{W^{2/3} \times V^4}{P_S} = \frac{16\,000^{2/3} \times 22^4}{13\,500}$$

Hence $A_C = 11\,018$ note much larger than before!

Worked example 8.7

A twin-screw vessel proceeds at a speed of 24 kt. She loses one of her propellers. Estimate her new forward speed.

For similar hulls, $A_C(1) = A_C(2)$ and $A_C \propto V^4$

Assume $W_1 = W_2$ or near enough, thereby cancelling each other out. Also $P_2 = (1/2)P_1$.

$$\frac{W^{2/3} \times V^4}{P_S} \text{ for condition (1)} = \frac{W^{2/3} \times V^4}{P_S} \text{ for condition (2)}$$

Hence $\dfrac{24^4}{P_1} = \dfrac{V_2^4}{(1/2)P_1}$ So $331\,776 \times \dfrac{1}{2} = (V_2)^4$

Thus $V_2 = 165\,888^{1/4}$ therefore $V_2 = 20.18$ kt   with one propeller.

Worked example 8.8

A twin-screw Passenger Cargo ship is of 19 470 tonnes displacement. She has the following particulars:

Ship speed (kt)	15	16	17	18
P_{NE} (kW)	2990	3750	4620	5640
QPC	0.730	0.730	0.720	0.710

(a) Determine the service speed for the vessel if the brake power of each engine is limited to 4050 kW (i.e. a total of 8100 kW).

Assume weather and appendage allowances = +30%.
Assume propeller shaft efficiency = 97%.

(b) Estimate the A_C corresponding to the obtained speed.

(a) $P_E = P_{NE} +$ (weather and appendage allowances)

$\qquad = P_{NE} \times (100\% + 30\%) = 1.30 \times P_{NE}$ at each speed in Table 8.1

$\qquad P_E/P_D = QPC$ Thus $P_D = P_E/QPC$ at each speed in Table 8.1

$\qquad P_D/P_B =$ Propeller shaft efficiency = 97%

Thus $P_B = P_D/0.97$ at each speed in Table 8.1

Fig. 8.2   Powering for Worked example 8.8.

Using the derived formulae and the given information it is possible to calculate the brake powers for ship speeds 15, 16, 17 and 18 kt. The results are shown in Table 8.1.

The brake powers were then plotted against the ship speeds. This is shown in Figure 8.3.

At the intersection of the curve with the total limiting brake power of 8100 kW, the speed was found to be 16.75 kt.

Thus the ship's service speed is 16.75 kt.

(b) $A_C = \dfrac{W^{2/3} \times V^3}{P_B} = \dfrac{19\,470^{2/3} \times 16.75^3}{8100}$

Thus $A_C = 420$ @ V of 16.75 kt.

Table 8.1   Powers against ship speed for Worked example 8.8.

Ship speed (kt)	15	16	17	18
P_{NE} (as given)	2990	3750	4620	5640
$P_E = 1.3 \times P_{NE}$	3887	4875	6006	7332
$P_D = P_E/QPC$	5325	6678	8342	10327
$P_B = P_D/0.97$	5490	6885	8600	10646

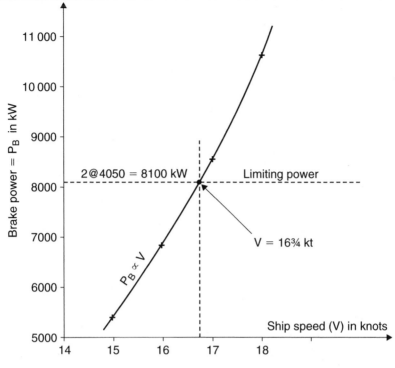

Fig. 8.3 $P_B \propto V$ for Worked example 8.8.

Worked example 8.9

Towing tank tests for a ship showed that when extrapolated to full size ship the P_{NE} values at certain speeds would be as shown in the following table:

Ship speed (kt)	14	15	16	17	18
P_{NE} (kt)	3540	4500	5500	6800	8450

(a) Draw a graph of P_{NE} against ship speed.
(b) Using the following data, determine the minimum shaft power (P_S) required to give a service speed of 16.50 kt: appendage allowances = 5%, weather allowances = 10%, shaft efficiency = 97%, QPC = 0.720.
(c) Calculate the corresponding A_C if this ship's displacement is 32 728 tonnes.

From Figure 8.4, it can be observed that when V is 16.5 kt, the P_{NE} is 6100 kW.

(b) $P_E = P_{NE} +$ (weather and appendage allowances)

$P_E = (100\% + 10\% + 5\%) \times P_{NE}$

$P_E = 1.15 \times P_{NE} = 1.15 \times 6100 = 7015\,\text{kW}.$

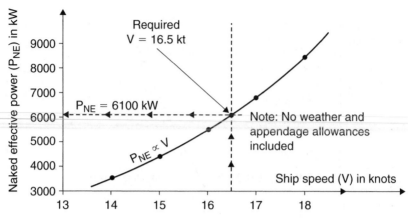

Fig. 8.4 $P_{NE} \propto V$ for Worked example 8.9.

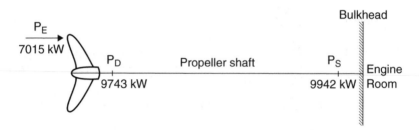

Fig. 8.5   Power values for Worked example 8.9.

$$P_E / P_D = QPC = 0.720 \quad \text{Thus} \quad P_D = P_E / 0.720 = 7015 / 0.720$$

Hence $P_D = 9743\,\text{kW}$

$$P_D / P_S = \text{Shaft efficiency} = 98\% \quad \text{So} \quad P_S = P_D / 0.98 = 9743 / 0.98$$

Hence $P_S = 9942\,\text{kW}.$

Check: $P_S = \dfrac{P_E}{QPC \times \text{shaft efficiency}} = \dfrac{1.15 \times 6100}{0.720 \times 0.98}$

$P_S = 9942\,\text{kW}$ (as before)

(c) $A_C = \dfrac{W^{2/3} \times V^3}{P_S} = \dfrac{33\,728^{2/3} \times 16.50^3}{9942}$

Hence $A_C = 462$ @ V of 16.5 kt.

Questions

1 For a new vessel, the effective power is 3479 kW, delivered power is 4785 kW and the brake power is 5178 kW. Calculate the QPC and the PC.

2 A Bulk Carrier has a displacement of 60 144 tonnes, a service speed of 14.80 kt and a shaft power of 8738 kW. What is the service speed for a similar Bulk Carrier having a displacement of 62 250 tonnes with a shaft power of 8450 kW?

3 A Passenger Liner has a displacement of 28 333 tonnes, a speed of 28 kt and a brake power of 9000 kW. Using the high service speed formula, estimate her A_C value.

4 A RO-RO vessel proceeds at a speed of 25 kt. She is a twin-screw ship. If she suddenly loses one of her propellers, what is her new forward speed? Clearly list any assumptions you make in your estimation.

5 A vessel has an LBP of 145 m and a service speed of 15.25 kt. Calculate the approximate value for A_C. State with reasoning if this propulsion machinery is very efficient, medium or of poor design.

6 A Very Large Crude Carrier (VLCC) has an input power of 26 500 kW within the Engine Room. Mechanical efficiency is 88.75%, propeller shaft losses are 2.65% and QPC is 0.621. Estimate the effective power (P_E) generated at the propeller tips, after accounting for all power losses enroute.

Chapter 9

Preliminary design methods for a ship's propeller and rudder

Propeller design

Consider first a method for obtaining the characteristics for a new propeller. There are several methods for determining the pitch and diameter of a propeller. Those most used are via model propeller experiments carried out in a cavitational tunnel. These tests produce a propeller chart known as a B_u or a B_p chart.

B_u charts use a thrust power constant based on P_T located at the propeller tips. B_p charts use a delivered power constant based on P_D located at the propeller's tail shaft (Figure 9.1).

Figure 9.1 shows a typical B_p chart for a ship's propeller. On this chart the following can be seen: B_p curves, pitch ratio lines, propeller efficiency curves, slip constant 'δ' curves and an optimum pitch ratio curve. The most efficient of all the propeller designs employs the use of this optimum pitch ratio line. This is shown in Worked example 9.1.

This particular B_p chart was for a 4-bladed propeller having aerofoil shaped blades. Other charts or diagrams will be specifically for 2-, 3-, 4-, 5-, 6- or 7-bladed propellers. Other diagrams will be for pear shaped blades, for round back blades or for segmental shaped blades.

Each B_p chart will have a specific value for the blade area ratio (BAR) (see Figure 9.2). Depending upon the design, this BAR value can range from being 0.20 to being 1.25. The propellers in this B_p chart all have a specific value for the BAR of 0.45.

Each B_p chart will also have a specific value for the thickness fraction (t_f) (see Figure 9.2). Generally 't_f' will range between 0.04 and 0.05. The propellers in the shown B_p chart all have a specific value for the 't_f' of 0.045.

Figure 9.2 illustrates these parameters. At this stage the reader will already be aware of the complexity of propeller design. There are many things to consider and many variables to take into account.

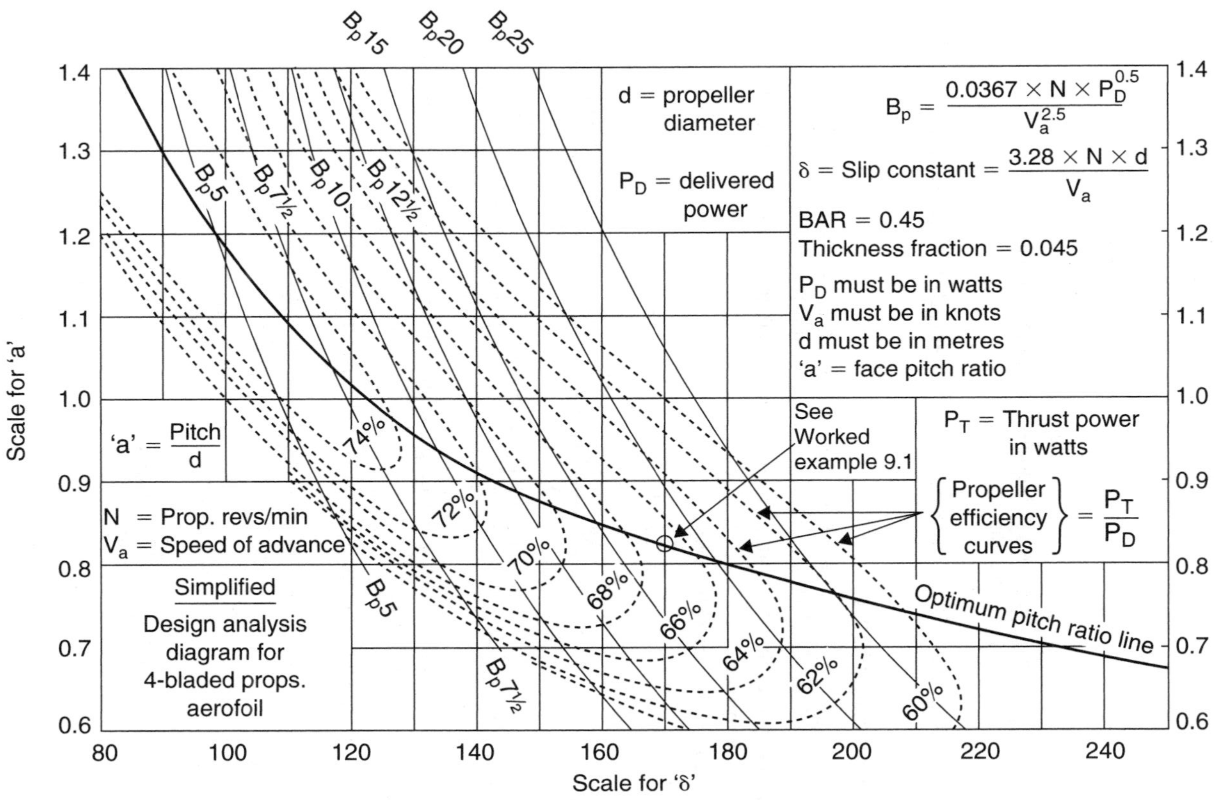

B_p15
B_p20
B_p25
Scale for 'a'
1.4
1.3
1.2
1.1
1.0
0.9
0.8
0.7
0.6
B_p5
$B_p7\frac{1}{2}$
B_p10
$B_p12\frac{1}{2}$
d = propeller diameter
P_D = delivered power
$B_p = \frac{0.0367 \times N \times P_D^{0.5}}{V_a^{2.5}}$
δ = Slip constant = $\frac{3.28 \times N \times d}{V_a}$
BAR = 0.45
Thickness fraction = 0.045
P_D must be in watts
V_a must be in knots
d must be in metres
'a' = face pitch ratio
'a' = $\frac{\text{Pitch}}{d}$
74%
72%
70%
68%
66%
64%
62%
60%
See Worked example 9.1
P_T = Thrust power in watts
Propeller efficiency curves = $\frac{P_T}{P_D}$
N = Prop. revs/min
V_a = Speed of advance
Simplified
Design analysis diagram for 4-bladed props. aerofoil
Optimum pitch ratio line
80
100
120
140
160
180
200
220
240
Scale for 'δ'

Fig. 9.1 A typical B_p propeller chart or diagram.

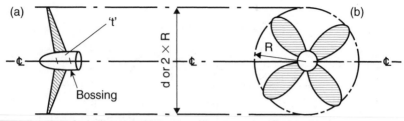

$$\text{Thrust deduction fraction} = \frac{\text{'t'}}{\text{d}} \quad \text{'t' measured at centreline } \mathcal{C}$$

$$\text{BAR} = \frac{\text{Surface area of blades}}{\text{Propeller disc area}} = \frac{\text{Shaded area}}{\pi \times d^2 \times \frac{1}{4}}$$

Fig. 9.2   Blade thickness fraction and BAR parameters.

Worked example 9.1

For a ship the delivered power is 4750 kW, propeller rpm are 100 and the velocity of advance (V_a) is 11.5 kt. Calculate the B_p constant, the propeller efficiency, the slip constant 'δ', the propeller pitch, the pitch ratio and the propeller diameter.

Using the B_p chart in Figure 9.1:

$$P_D = 4\,750\,000\,W, \quad N = 100\,\text{rpm}, \quad V_a = 11.5\,\text{kt}$$

$$B_p = \frac{0.0367 \times N \times P_D^{0.50}}{V_a^{2.5}} = \frac{0.0367 \times 100 \times 4\,750\,000^{0.50}}{11.5^{2.5}}$$

Hence B_p constant $= \dfrac{0.0367 \times 100 \times 2179}{448.5} = 17.83$

At the intersection of the optimum pitch ratio curve and the B_p value of 17.83, it can be observed that:

Propeller efficiency $= 66.7\%$ $\delta = 170$ Pitch ratio 'a' $= 0.825$

$$\delta = \frac{3.28 \times N \times d}{V_a} \quad \text{So} \quad d = \frac{\delta \times V_a}{3.28 \times N}$$

Hence $d = \text{Propeller diameter} = \dfrac{170 \times 11.5}{3.28 \times 100} = 5.96\,\text{m}$

$$a = \frac{\text{Pitch}}{\text{Diameter}} = 0.825$$

So   Propeller pitch $= 0.825 \times 5.96 = 4.92\,\text{m}$

Procedure steps

1. Evaluate the B_p constant, i.e. 17.83.
2. Track down the B_p curves at a value of 17.83 until an intersection is made with the optimum pitch ratio line. This will give the best propeller design.

3. At the intersection point, determine by visual proportion the propeller efficiency, the slip constant and the pitch ratio values, i.e. 66.7%, 170 and 0.825 respectively.
4. Transpose the slip constant formula to obtain the propeller diameter, i.e. 5.96 m.
5. Transpose the pitch ratio formula and obtain the propeller pitch, i.e. 4.92 m.

Worked example 9.2

Using the answers obtained in Worked example 9.1, proceed to calculate the thrust power (P_T), thrust in kN, blade area in m^2, thrust in kN/m^2 and the blade thickness in metres at the centreline of the propeller bossing:

$$P_T / P_D = \text{Propeller efficiency} \quad \text{So} \quad P_T = P_D \times \text{Propeller efficiency}$$

Hence $\quad P_T = 4750 \times 66.7\% = 3168 \text{ kW}$

$$P_T = \text{Thrust} \times V_a \qquad \text{Thrust} = \frac{P_T}{V_a} = \frac{3168}{(11.5 \times 1852)/3600}$$

Hence $\quad$ Thrust = 535.5 kN

$$\text{BAR} = \frac{\text{Area of blades}}{\text{Propeller disc area}} = 0.45 \text{ (given on } B_p \text{ chart)}$$

$$\text{Area of blades} = 0.45 \times \left(\frac{\pi \times d^2}{4} \right) = \frac{0.45 \times 3.142 \times 5.96 \times 5.96}{4}$$

Hence $\quad$ area of blades = 12.56 m^2

$$\text{Thrust on propeller baldes} = \frac{\text{Thrust}}{\text{Blade area}} = \frac{535.5}{12.56}$$

$$= 42.64 \text{ kN/m}^2$$

If this thrust of 42.64 kN/m^2 is too high, then imploding bubbles appear on the propeller blades. These imploding bubbles cause cavitation. Cavities in time will appear in the propeller blades. Metal is being literally sucked out of the propeller blade material thereby weakening the blade.

Prolonged cavitation at the tips of the propeller could cause cracking and eventually fracturing from the rest of the blade. This in turn causes imbalance of the propeller leading to propeller exited vibration problems.

At the design stage, this could necessitate going to another B_p chart with a higher BAR, say 0.50 or 0.60. Perhaps an increase in the propeller diameter or a change in the propeller revolutions might solve the problem.

Finally, the blade thickness fraction = t/d = 0.045 (given on B_p chart)

So blade thickness at propeller bossing = t = 0.045 $\times$ d = 0.045 $\times$ 5.96

$$= 0.268 \text{ m}$$

Fig. 9.3   A large modern container ship propeller nearing completion. This
70 tonnes screw has a diameter of 8550 mm and is designed to absorb
42.5 MW at 94 r/min.
Reference source: 'Developments in Marine Propellers' by Dr. G. Patience for
Institute of Mechanical Engineers (London) January 1991.

Summary remarks

A propeller with a larger diameter operating at lower revolutions (say
85 rpm as on Very Large Crude Carriers (VLCCs)) would give a larger propeller efficiency value. To increase the efficiency of a ship's propeller, a nozzle can be fitted. This nozzle may be of fixed design, as installed on large
vessels. It may be movable for smaller diameter propellers. For more
details on these nozzles, see Chapter 21.

Materials used in the construction of propellers are high tensile brass,
high manganese alloys, nickel–aluminium alloys, stainless steel, cast iron
and polymer plastics such as nylon and fibreglass.

Figure 9.3 shows a photograph of a 70 tonnes propeller. The heaviest finished weight for a propeller, up to September 2003 is 105 tonnes.

Propeller diameters can range from being 2 m up to 11 m (fitted on
VLCCs and Ultra-Large Crude Carriers (ULCCs)). See Figure 9.4 which is a
photograph of an 11 m diameter propeller.

Fig. 9.4   The economy propeller, with a finished weight of 69 tonnes and a diameter of 11 m, before retrofit to a 320 ktonnes dwt ULCC. *Reference source*: 'Developments in Marine Propellers' by Dr G. Patience for Institute of Mechanical Engineers (London) January 1991.

Rudder design

The profile of a rudder is obtained by calculating the value A_R (see Figure 9.5).

$$A_R = K \times LBP \times d \quad (m^2) \qquad \text{for single screw and twin screw ships}$$

where:

K = a coefficient dependent upon the type of ship (see Table 9.1),
LBP = length between perpendiculars in metres,
   d = fully loaded draft in salt water, i.e. the Summer Loaded Waterline (SLWL).

Generally the faster ships in their own class will have comparatively lower K values and hence smaller size rudders.

Rudders are generally designed to perform one of two main functions, which is as follows:

1. To keep the ship on a straight line, from Port 'A' to Port 'B.' In other words, to have good course keeping properties.
2. To turn the ship in a small turning circle diameter. In other words, to have good turning characteristics, say in confined waters.

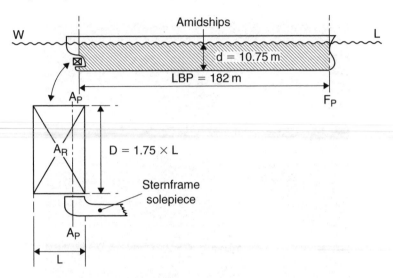

Fig. 9.5   Simplex balanced rudder with ship profile.

Table 9.1   K values for several Merchant ships

Type of ship	Typical value for K (%)
Container ships and Passenger Liners	1.2–1.7
General Cargo ships	1.5
Oil Tankers and Bulk Carriers	1.7
Lake steamers	2.00
Cross Channel ferries (RO-RO ships)	2–3
Coastal vessels	2.0–3.3
Tugs, Pilot vessels	2.5–4.0

For good course keeping and good turning properties, adhere to the K values in Table 9.1. To further increase the turning characteristics, simply increase the K value to obtain a larger rudder profile. However a larger rudder profile will decrease a ship's course keeping properties. Thus each ship type will need an appropriate K value and consequent appropriate size of rudder.

Worked example 9.3

A Bulk Carrier is to be fitted with a rectangular Simplex Balanced rudder. The rudder depth is to be 1.75 times the rudder length. LBP is 182 m with an SLWL of 10.75 m. Using Table 9.1, evaluate the area A_R. Proceed to determine the length and depth for this rudder.

$$A_R = K \times LBP \times SLWL = 1.7\% \times 182 \times 10.75 = 33.26\,m^2$$

Now $A_R = 33.26 = L \times D = 1.75 \times L^2$ So $33.26 = 1.75 \times L^2$

Thus $L = 4.36\,m$, $D = 1.75 \times L = 1.75 \times 4.36 = 7.63\,m$

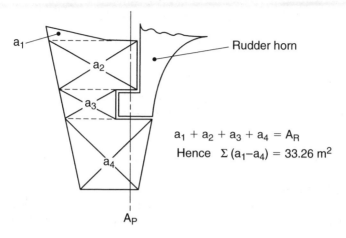

Fig. 9.6   A mariner rudder profile.

$a_1 + a_2 + a_3 + a_4 = A_R$
Hence $\Sigma (a_1-a_4) = 33.26 \ m^2$

Hence   Rudder area $= 33.26 \ m^2$, Rudder length $= 4.36 \ m$,
Rudder depth $= 7.63 \ m$.

If the selected rudder had been a Mariner type design then the various areas would have been added together to obtain the previously calculated A_R value of $33.26 \ m^2$. This is illustrated in Figure 9.6.

Size of the steering gear machinery for turning the rudder

Classification Society rules stipulate that the ship's steering gear must be capable of moving the rudder from 35° helm on one side to 30° helm on the other side, in 28 sec with the ship at full speed.

Most rudders have maximum rudder helm of 35° Port to 35° Starboard. Beyond this, the flow of streamlines around the rudder breaks down. Stalling occurs and the rudder efficiency deteriorates.

Linked with the size of the steering gear machinery are the forces on the rudder F, F_t and F_n. These are shown in Figure 9.7.

F_n = The rudder force normal to the plane of the rudder.
F_t = The transverse rudder force.
α = Rudder helm in degrees.

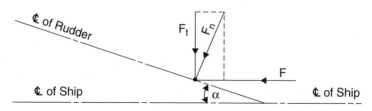

Fig. 9.7   Forces F, F_t and F_n on a ship's rudder.

Using Figure 9.7 $F_t = F_n \cos\alpha = F \sin\alpha \cos\alpha$ N

$F = \beta \times A_R \times V^2$ N   Acting parallel to the centreline of the ship

where:

V = ship speed in m/s,

β = 570–610, when the ship is in salt water, of modern rudder shape and a typical rudder helm up to 35° P&S maximum.

Grouping these equations together results in:

$F_t = \beta \times A_R \times V^2 \times \sin\alpha \cos\alpha$ N   the transverse rudder force

Worked example 9.4

Calculate the transverse force F_t on the rudder considered in the previous example when the rudder helm is 35° with a ship speed of 14.8 kt. Assume that $\beta = 580$.

$$\text{Ship speed} = V = \frac{14.8 \times 1852}{3600} = 7.61 \text{ m/s} \qquad A_R = 33.26 \text{ m}^2$$

$F_t = \beta \times A_R \times V^2 \times \sin\alpha \cos\alpha = 580 \times 33.26 \times 7.61^2 \times \sin 35° \cos 35°$

$F_t = 580 \times 33.26 \times 57.912 \times 0.5736 \times 0.8192$

$F_t = 524\,951$ N

$F_t = 525$ kN

Questions

1 List the variables associated with a propeller's B_p chart.
2 List the procedures steps for obtaining the propeller's pitch and diameter when using a B_p chart.
3 With the aid of sketches, describe exactly what BAR and blade thickness fraction signify.
4 Give the formula for:
   (a) delivered power B_p constant, (b) slip constant 'δ', (c) pitch ratio 'a' and (d) propeller efficiency.
5 For a new ship, the delivered power is 4500 kW, propeller revolutions are 98 rpm, velocity of advance is 12 kt. Calculate the B_p constant, propeller efficiency, slip constant 'δ', propeller pitch, pitch ratio and the propeller diameter. Use the B_p chart in Figure 9.1.
6 For a simplex balanced rectangular rudder, the rudder depth is to be 1.694 times the rudder length. The ship is 170 m LBP with an SLWL of 9.0 m and a 'k' value of 2%. Calculate the A_R, the rudder length and the rudder depth.
7 For a new ship, β is 585 and the A_R is 35.07 m². Calculate the transverse rudder force F_t, when the rudder helm is 35° when the ship is operating at 14 kt speed.

Nomenclature for ship design and performance

This glossary should be cross-referenced with the index to obtain a fuller understanding of each term.

2NH	a vibrating beam or ship with two points of zero magnitude in a horizontal mode
2NV	a vibrating beam or ship with two points of zero magnitude in a vertical mode
A_C	Admiralty Coefficient, connecting displacement, ship speed and power at the thrust block
Air Draft	a vertical distance measured from the waterline to the topmost structure of the ship
Alexander's formula	a formula linking the C_B, the ship speed and the ship's length between perpendiculars (LBP)
amidships	a position midway between the Aft Perpendicular (AP) and the Forward Perpendicular (FP)
amplitude	the distance or movement from an initial position of a vibrating beam
antinode	the point where the magnitude is the greatest for a vibrating beam
aperture	locality in which the propeller revolves
Apparent Slip	the difference between theoretical speed and actual speed of the ship
appendage	a small addition to the main part or main structure
A_R	area of the rudder in profile view
asymptotic water depths	water depths in which velocities, squats, propeller rpm, vibration frequencies, etc. remain unchanged
auxiliary machinery	machinery other than the ship's main engine
balance of weights table	a table linking the steel weight, wood and outfit weight and the machinery weight together
bale capacity	a capacity reflecting the stowage of bales or boxes in a hold or tween deck
basis ship	a ship very similar in type, size, speed and power to that of the new design
Becker twisted rudder	a rudder where the centreline at the top part is directionally different to the centreline at the lower part

(Continued)

blade area ratio	total area of blades divided by the propeller disc area
blade thickness fraction	blade thickness 't' divided by the propeller diameter
blockage factor	area of ship's midship section divided by the cross-sectional area of a river or canal
BM_T	value of transverse moment of inertia of waterplane/volume of displacement
bollard pull (max)	carried out on a tug to measure the pulling power when operating at maximum input of engine power
boot-topping	the vertical distance between the lightdraft and the summer load waterline
Bp chart	used to calculate the propeller efficiency, pitch ratio and propeller diameter
BSRA	British Ship Research Association
bulbous bow	fitted in the Fore Peak Tank to help reduce wave making resistance
bulbous stern	fitted in the Aft Peak Tank to help reduce wave making resistance
Bulk Carriers	workhorse vessels, built to carry such cargoes as ore, coal, grain, sugar, etc. in large quantities
Cargo–Passenger ship	a vessel that carries cargo and up to 12 paying passengers
cavitation	imploding bubbles appearing and collapsing on a propeller, causing metal to be drawn from these structures
C_B	block coefficient: linking the volume of displacement with LBP, Breadth Mld and Draft
CCC	confined channel conditions
C_D	deadweight coefficient: linking the deadweight and the displacement of a ship
CNC programme	Computer Numerical Cutting programme, using negative plates or computer tapes
computer packages	packages for estimating stability, trim, end drafts, shear forces and bending moments for a condition of loading
confined channel	a river or canal where there is a nearby presence of banks
contra-rotating propellers	one propeller fitted aft of another, one revolving clockwise and the other revolving anti-clockwise
Crash-stop manoeuvre	performed on Ship Trials to measure stopping distances and times with rudder helm is fixed at zero
C_W	waterplane area (WPA) coefficient: linking the WPA with the LBP and the ship's Breadth Mld

d, H or T	draft of ship
DfT	Department of Transport
deadweight	the weight that a ship carries
deck camber	transverse curvature of a deck, measured from deck height at centreline to deck height at side at amidships
deck sheer	longitudinal curvature of a deck, measured vertically from amidships to the deck at AP or FP
Depth Moulded	measured from top of keel to underside of uppermost continuous deck, at amidships
D_h	depth of hold that contains cargo
Dieudonne spiral	a manoeuvre that measures course heading response to a requested rudder helm, usually on Ship Trials
displacement	for all conditions of loading, it is the lightweight plus any deadweight
domain of ship	mainly the area in which the pressure bulbs exist on a moving vessel
Draft Moulded	distance from the waterline to the top of keel, measured at amidships
D_t	depth of tanks containing cargo oil and water ballast capacity
dwt	abbreviation for deadweight
engine efficiency	the link between the power measured at the thrust block and the input power within the engine room
entrained water	water picked up by a moving ship due to frictional contact with the hull of the ship
even keel	a vessel with no trim: where the Aft draft has the same value as the Forward draft
F_B	breadth of influence in open water conditions
F_C	fuel coefficient for estimating the fuel consumption/day in tonnes
F_D	depth of influence, located at the tailshaft of the propeller shaft
flume	a tunnel of moving water used to measure the total drag or resistance on suspended stationary ship models
forced vibration	vibration eminating from machinery within or on the ship
Froude Number	a number without units, linking the speed, gravity 'g' and the ship's LBP
fuel cons/day	fuel consumption per day usually measured or recorded in tonnes/day

(Continued)

General Particulars	LBP, Breadth Mld, Depth Mld, Draft Mld, lightweight, deadweight, displacement, C_B, service speed, etc.
geosim	short for 'geometrically similar' ship or ship model
GM_T	transverse metacentric height
GPS	Global Positioning System for tracking a ship using satellite navigation techniques
grain capacity	grain volume, about 1.5% less than the moulded volume
Grim vane wheel	a wheel fitted aft of the propeller, to increase propulsion efficiency
H	water depth
hull efficiency	a link between the effective power and the thrust power
hull weight	steel weight plus the wood and outfit weight
hydrostatic curves	used for calculating the trim and stability values for various conditions of loading
IMO	International Maritime Organisation
inertia coefficient	used for obtaining moments of inertias of waterplanes
insulated volume	used for capacities on refrigerated cargo ships
Interaction	action and reaction of ships when they get too close to one another or too close to a river bank
invoice steel weight	weight purchased by the shipyard
Kappel propeller	a new propeller concept
KB	vertical centre of buoyancy
KG	vertical centre of gravity
KM	height of Metacentre above base
K_{max}	coefficient for estimating maximum ship squat
K_{mbs}	mean bodily sinkage coefficient, when dealing with ship squat
knot	1852 m/h
$K_{o/e}$	coefficient used, when calculating squat at opposite end of ship to where maximum squat occurs
Kort nozzle	a fixed annulus of steel around a propeller to make it more efficient
Kort rudder	a moveable annulus of steel around a propeller to make it more efficient
K_t	trim coefficient, when dealing with ship squat

Term	Definition
LCB	longitudinal centre of buoyancy
LCF	longitudinal centre of flotation
L_h	total length of holds containing cargo
lightship draft	draft of ship when ship is empty, with a deadweight of zero
lightweight	empty ship, with boilers topped up to working level
LNG ships	liquified natural gas carrier with cargo at $-161°C$
LOA	length overall, from the foremost part of the bow to the aftermost part of the stern
longitudinal	running from bow to stern in a fore and aft direction
LPG ships	liquified petroleum gas carrier with cargo at $-42°C$
L_t	the length of the ship carrying cargo oil and water ballast
machinery weight	a total of the main engine weight plus all the auxiliary machinery
Mariner rudder	a rudder that is connected to a rudder horn and does not have a sternframe solepiece support
MCO	Maximum Continuous Output power
MCR	Maximum Continuous Running power
MCTC	moment to change trim 1 cm
measured mile	used on Ship Trials and is 1852 m in length
mode	type of vibration, for example vertical, horizontal or torsional
moulded capacity	calculated using only the moulded dimensions of a hold or tween deck
MP 17 ship	a 17000 tonnes dwt multi-purpose vessel, with a 17-person crew and using only 17 tonnes of fuel/day
natural vibration	vibration attributable to natural forces like wind, waves and gravitational effects
nautical mile	1852 m
net scantling weight	weight actually ordered by the shipyard
net steel weight	weight that ends up in the new ship
Nm	mean revolutions per nautical mile
node	a point on a vibrating beam where the amplitude is zero
NPL	National Physical Laboratory

(Continued)

Term	Definition
open water	a stretch of water where there are no adjacent river or canal banks
out to out	the total movement of a vibrating plate
OWC	open water conditions
P&S	Port and Starboard
Panamax vessel	a vessel having a Breadth Mld no more than 32.26 m
Passenger Liners	vessels travelling between definite Ports, with timetabled departure and arrival dates
Passenger–Cargo ship	a vessel that carries cargo and more than 12 paying passengers
P_B	brake power for diesel machinery
PBCF	Propeller Boss Cap Fins
PC	propulsive coefficient
P_D	delivered power, located at the tailshaft
P_E	effective power at the propeller tips
pitch	the distance moved forward by a point on the propeller after one complete revolution
pitch ratio	links the propeller pitch and the propeller diameter
P_{NE}	naked effective power, without any weather and appendage allowances
pods	fixed or azimuth power units suspended and operated from the steering gear compartment
Port	left side of a ship when looking forward
pressure bulbs	bulbs of force that build up around a moving vessel and disappear when vessel stops
propeller efficiency	links the thrust power with the delivered power
propeller shaft efficiency	links the delivered power with the power measured at the thrust block
P_S	shaft power for Steam Turbine machinery
P_T	thrust power located at the propeller tips
QPC	quasi-propulsive coefficient: linking the effective power with the delivered power
raked propeller blades	where the propeller blades are raked aft to improve the aperture clearances
real slip	theoretical speed minus the speed of advance
reefer ships	abbreviation for refrigerated cargo ships
resonance	where two vibration frequencies have the same value, thereby causing vibration problems

Term	Definition
retrofits	something added to or deducted from a ship, after inefficiencies occurred with the original design concept
R_f	frictional resistance
RO-RO Ships	roll–roll vessels that carry cars/lorries and passengers
RPNM	revolutions per nautical mile
R_r	residual resistance
R_T	total resistance
rudder fins	fins welded to the rudder to improve propulsion efficiency
rudder helm	angle to which a rudder is turned, with the maximum angle normally being 35° P&S
rudder horn	the support for a Mariner design rudder
rudder stock	connects the top of rudder to the steering gear machinery, used for turning the rudder
scantlings	measurements or dimensions of a plate
Schilling rudder	type of rudder for improving ship-handling performance
shallow water	where the depth of water reduces the ship speed and propeller revolutions, increases squat, reduces rolling motions, etc.
ship surgery	lengthening, deepening or widening a ship after cutting her transversely, along or longitudinally
shot-blasted	a procedure in the shipyard for cleaning the steel plates prior to giving them a primer coat paint
sighting posts	used on Ship Trials, they are spaced on shore 1852 m apart
Simplex Balanced rudder	a rectangular-shaped rudder that is supported at its base by a sternframe solepiece
slack water	where the tide has zero speed or direction
slip constant	links the propeller revolutions, the propeller diameter and the velocity of advance
SLWL	summer load waterline, similar to Draft Mld
spade rudder	a rudder held only by its rudder stock, with no rudder horn and no solepiece support
speed–length law	links ship speed and the LBP

(*Continued*)

Term	Definition
squat	loss of underkeel clearance as a ship moves forward from rest
St Lawrence Seaway max	where the Breadth Mld is not to exceed 23.8 m
Starboard	right side of a ship when looking forward
STAT 55	standard design 55 000 tonnes dwt Oil Tanker
stern tunnel	structure fitted over the propeller to improve propulsion characteristics
sternframe solepiece	bottom support of a ship's sternframe
Supertankers	similar to a VLCC: having a dwt of 100 000–300 000 tonnes
synchronisation	where two vibration frequencies have the same value, thereby causing vibration problems
tailshaft	the aftermost section of the propeller shaft
TCD	turning circle diameter
tee-duct	fitted in the Fore Peak Tank to help bring the vessel to zero speed
thrust block	takes all the forces produced by the propulsion system built into the ship
tonne	equivalent to 1000 kg
towing tanks	long tanks used by ship models for resistance tests
TPC	tonnes per centimetre immersion
transverse	running from Port to Starboard across the ship
transverse squat	caused by ships overtaking or passing in a river
transverse thruster	transverse fitted propeller to assist in moving vessel sideways towards or away from a jetty
Trial speed	because of lower displacement, usually 0.50–1.00 kilotonnes above the designed service speed
trim	the difference between the forward draft and the Aft draft
ukc	underkeel clearance
ULCC	ultra-large crude carrier, say over 300 000 tonnes dwt
V_a	velocity of advance
Vectwin rudder	twin rudder concept of recent years
VLCC	very large crude carrier, say 100 000–300 000 tonnes dwt
Voith-Schneider unit	a propulsion system that uses vertical revolving blades to good effect
V_s	ship speed

V_t	theoretical speed … or the total volume over the cargo network of tanks in an Oil Tanker
W&O weight	wood and outfit weight
wake speed	the difference between ship speed and velocity of advance
windlass	structure used for raising and lowering the anchor and anchor cable
WPA	waterplane area
WSA	wetted surface area
y_2	underkeel clearance when ship is moving ahead at a particular speed
y_0	original underkeel clearance when ship is stationary
Zig-zag manoeuvre	part of a Ship's Trial to examine rudder helm against a ship's course heading

Part 2
Ship Performance

Chapter 10
Modern Merchant ships

This chapter is a journalistic review of ships in operation at the present time. The first indicator for the size of a ship is usually the deadweight (dwt) measured in tonnes and the dwt is the weight a ship carries.

With some designs, like Passenger Liners, the size can be indicated by the Gross Tonnage (GT) measured in tuns. With Gas Carriers, it is usually the volume of gas carried, measured in cubic metres. The size of a Tug is gauged by the value of the bollard pull, measured in tonnes.

Ships are the largest moving structures designed and built by man. The following notes illustrate the characteristics relating to several ship types such as:

- Oil Tankers
- Product/Chemical Carriers
- OBOs and Ore Carriers
- Bulk Carriers
- General Cargo ships
- Gas Carriers
- Passenger Liners
- Container ships
- RO-RO ferries
- Tugs
- Hydrofoils/Hovercraft
- SWATH (Ship With A Twin Hull) designs.

Oil Tankers

These vessels may be split into three groups, namely the medium size tankers, the Very Large Crude Carriers (VLCCs) or Supertankers and the Ultra Large Crude Carriers (ULCCs). By definition, Oil Tankers are ships that carry liquid in bulk. They are slow-moving full-form vessels. Table 10.1 shows the Main Dimensions for these ships.

Since 1994 new tankers having a dwt of greater than 5000 tonnes are required to have a double hull construction in way of the Main Cargo network of tanks. Some shipowners have gone further in that they have requested a double-bottom construction beneath the main network of cargo tanks.

Table 10.1 Some characteristics of modern Oil Tankers (see also Chapter 11)

Type of ship	Typical dwt (tonnes)	LBP (m)	Br. Mld (m)	Typical C_B fully loaded	Service speed (kt)
Medium size	50 000–100 000	175–250	25–40	0.800–0.820	15.00–15.75
VLCCs and ULCCs	100 000–565 000	250–440	40–70	0.820–0.850	13–15.75

They are similar in effect to Container ships in way of their midship sections. They are now known as double-skin Tankers. It is hoped by having these tanker designs, the problems of oil pollution following collision or grounding are greatly decreased.

Product/Chemical Carriers

These vessels are tankers that do not carry crude oil. They are a spin off from the Oil Tanker design because they carry several different types of refined petroleum based products. It is a multi-type oil-carrying vessel. Economics feedback has shown that there is a profitable place in the shipping market for this hybrid design (see Chapter 11).

OBOs and Ore Carriers

OBOs are Oil and Ore carriers. They carry oil in their wing tanks and iron ore in their centre tanks, but *not* on the same voyage. Ore Carriers convey iron ore in their holds that stretch right across the width of the vessel. They are slow-moving full-form vessels. Table 10.2 shows the Main Dimensions for these ships.

Table 10.2 Some characteristics of modern OBOs and Ore Carriers

Type of ship	Typical dwt (tonnes)	LBP (m)	Br. Mld (m)	Typical C_B fully loaded	Service speed (kt)
OBO ships	Up to 173 000	200–300	Up to 45	0.780–0.800	15.00–16.00
Ore Carriers	Up to 322 000	200–320	Up to 58	0.790–0.830	14.50–15.50

Bulk Carriers

These are vessels that carry cargoes such as grain, iron ore, sugar, bauxite or coal. They are slow-moving full-form ships. The smaller and older Bulk Carriers have C_B values of 0.750–0.800. The newer and larger generation of this ship type have C_B values of 0.800–0.830. They are very popular on the Great Lakes in Canada where they are known as 'Great Lakers'. Service speeds generally range from 15.00 to 15.75 kt (see Chapter 11).

General Cargo ships

As the name implies, these vessels carry all sorts of general cargo, the main one being the carriage of grain. Sometimes known as 'Cargo Tramps', they go anywhere, carry anything, carry anybody, at any time ... provided the price is right!!

Their dwt ranges from 3000 to 15 000 tonnes, their length between perpendiculars (LBPs) from 100 to 150 m and their breath moulded (Br.Mld) from 15 to 25 m. When fully loaded, their C_B can be 0.675 up to 0.725 with service speed in the range of 14–16 kt. They are slow medium-form vessels (see Chapter 11). They may be of refrigerated design, known in the shipping industry as 'Reefers'.

Gas Carriers

In principle, the design is 'a box within a box that is separated by a void space', similar in effect to the principle of a flask. Gas Carriers can be split into two distinct groups. One is the liquefied natural gas (LNG) carrier. The other is the liquefied petroleum gas (LPG) carrier.

LNG is mainly methane and ethane. LNG ships carry their cargo at $-161°C$, at a relative density of approximately 0.600 with a volume contraction ratio of 1 in 600. LNG cargo is carried at ambient pressure.

LPG is mainly propane and butane. LPG ships carry their cargo at $-42°C$, at a relative density of approximately 0.500 with a volume contraction ratio of 1 in 300. LPG cargo may be carried under pressure.

The cargo tank construction of LNG and LPG ships can be of (a) prismatic design (b) membrane design or (c) spherical design. Materials used for these cargo tanks can be aluminium, balsa wood, plywood, invar or nickel steel, stainless steel, with pearlite and polyurethane foam.

Because of the demand for insulation at these extremely low cargo temperatures, the first cost of these specialised ships are extremely high. A very high standard of workmanship is required for the building of these types of vessel.

Their capacity ranges from 75 000 to 138 000 m³ of gas, their LBPs up to 280 m and their Br. Mld from 25 to 46 m. When fully loaded, their C_B can be 0.660 up to 0.680 with service speed in the range of 16–20.75 kt. They are fine-form vessels (see Chapter 11).

Passenger Liners

These are normally 'one-off' special ships. As well as passengers (up to 3000+), they carry cargo and cars from Port to Port on a regular timetable. They have been likened to 'floating villages' (see Chapter 11).

Their Gross Tonnage can be up to 150 000 tuns (QM2 in 2003), their LBPs from 200 to 345 m (QM2) and their Br. Mld from 20 to 48 m. When fully loaded, their C_B can be 0.600 up to 0.640 with service speed in the range of 22–30 kt. They are fast fine-form vessels.

Container ships

In principle they are 'boxes or containers within a box'. These boxes or containers have dimensions of 2.60×2.45 m with lengths of 6.10, 9.15 and 12.20 m. Containers are made in steel, aluminium or GRP. They are also of refrigerated design, thus advantageous for long voyages between Australia or New Zealand and the UK.

Because the cargo is put into containers there are several immediate advantages:

- The cargo can be loaded and discharged much faster than for General Cargo ships. Hence, less time is spent in Port.
- Consequently, less Port dues are paid by the shipowner.
- More voyages per year, hence more income for the shipowner.
- Less pilferage, so lower insurance costs for the shipowner.
- When compared to General Cargo ships, less number of Crew are required on these ships.
- They are usually larger and faster than General Cargo ships.

The track record shows that a container can generally be loaded or discharged every 3 min. Some Container ships come in on the morning tide, discharge, reload and sail out on the next tide.

Their dwt ranges from 10 000 to 72 000 tonnes, their LBPs from 200 to 300 m and their Br.Mld from 30 to 45 m. When fully loaded, their C_B can be 0.560 up to 0.600 with service speed in the range of 20–28 kt. They are fast fine-form vessels (see Chapter 11). Because of increases in oil fuel costs, the service speed for new orders for Container ships are at present in the 20–22 kt range.

RO-RO vessels and ferries

The design concept of a 'roll on–roll off' vessel is that of a moving multistorey car park. They carry cars, lorries, trailers, cargo and passengers. They may be single-screw or twin-screw design. They are ideal for short fast trips across shipping lanes such as the English Channel, North Sea and the Irish Sea.

The cars and trailers can be driven into the ship at the stern, at the bow or through several side openings along the length of the ship. Following recent several disasters with these vessels, entering the ship via the bow has become less popular for new orders for these vessels.

On this point of disasters, it should be remembered that RO-RO ferries have a history that after being involved in a collision with another vessel, they can capsize and sink in *only 1.50 min*. This is mainly due to the very wide spaces athwartships.

Their dwt ranges from 2000 to 5000 tonnes, their LBPs from 100 to 180 m and their Br.Mld from 21 to 28 m. When fully loaded, their C_B can be 0.550 up to 0.570 with service speed in the range of 18–24 kt. They are fast fine-form vessels (see Chapter 11).

Tugs

These have been called 'the tractors' of the shipping industry. There are several types of Tugs. They can be an Ocean-going and Salvage Tug, a Coastal Tug, a Port Tug or an Inland waterway Tug. Tugs can be single screw or twin screw. LBP/Br.Mld is of the order of 3:1 whilst Br.Mld/Draft Mld averages out at 2.35:1.

They can be driven by a conventional propeller (with a Kort rudder) on the end of a horizontal shaft. They can also be powered by a Voith–Schneider (VS) design or by azimuth (ASD) propulsion units. Table 10.3 indicates some characteristics of modern Port Tugs.

Table 10.3 Characteristics of modern Port Tugs and escort speed can be up to 12 kt

Type of ship	Bollard pull (tonnes)	LBP (m)	Br. Mld (m)	Draft Mld (m)	Brake power (kW)
Port Tug	40–115	26–47	9–16	4–7	2835–7385

In his book on Tugs, Captain Hensen produced several graphs (page 82) and a table (page 149) for average bollard pull. In 2003, the author of this book plotted this information and converted it to metric units. The revised research work of Captain Hensen now appears as:

For ASD designed Tugs, bollard pull = $0.016 \times P_B$ tonnes
for bollard pulls >50 tonnes

For VS designed Tugs, bollard pull = $(0.012 \times P_B) + 7$ tonnes
for bollard pulls >40 tonnes

For Oil Tankers and large Bulk Carriers awaiting the assistance of Tugs, it is possible to estimate the average total bollard pull required:

$$\text{Total bollard pull required} = \left(60 \times \frac{W}{100\,000}\right) + 40 \text{ tonnes}$$

Captain Hensen (1997).

where W = ship's displacement in tonnes.

If in bad weather or in shallow waters, an *addition* is made. If in calm weather or in deep waters, a *reduction* is made to this average total bollard pull. This addition or reduction is based on the work experience of the Tug's master but it could be up to ±50 tonnes.

Worked example 10.1

What is the total bollard pull required for a VLCC of 100 000 dwt and 122 500 tonnes displacement?

$$\text{Total bollard pull required} = \left(60 \times \frac{W}{100\,000}\right) + 40\text{ tonnes}$$

$$= (60 \times 122\,500) + 40 = 113.5\text{ tonnes}$$

If the displacement is not available, one can use another approximation for the required bollard pull. This time it is associated with the LBP.

For Oil Tankers and Bulk Carriers awaiting the assistance of Tugs:

Total bollard pull required = $(0.7 \times \text{LBP}) - 35$ tonnes
Hensen graphs (1997) and Barrass equations (2003)
Applicable for LBP > 140 m

Worked example 10.2

A VLCC is 214 m LBP. Estimate the total bollard pull required for this vessel.

$$\text{Total bollard pull required} = (0.7 \times \text{LBP}) - 35\text{ tonnes}$$

$$= (0.7 \times 214) - 35 = 115\text{ tonnes}$$

Tugs are highly manoeuvrable, comparatively fast for their length and have a C_B of the order of 0.500–0.525. As well as towing, some special Tugs known as Push Tugs move dumb barges along rivers. Tugs can also be used for fire-fighting and pollution control duties.

Hydrofoils/Hovercraft

Hydrofoils are small craft that rise out of the water (on vertical foils) at the bow when at high speed. This speed can be as high up to 40 kt. Hovercraft rise completely out of the water at high speed on a pressure cushion. Again the speed can be as high as 40 kt. Both the Hydrofoil and the Hovercraft are ideal for short fast trips across shipping lanes such as the English Channel, fiords, North Sea and the Irish Sea.

SWATH designs

These are Small Waterplane Area-Twin Hull designs or Ships With A Twin Hull. The SWATH has two separate hull form joined together by a horizontal bridge type construction. They are transversely very stable, with low noise levels. They have low vibration problems and superior sea-keeping qualities.

It has been claimed for the *Radisson Diamond* (a SWATH design of 116 m LBP and 32 m Br.Mld) that the hull roll is only 20% of that of a mono-hull design. The best-known one at the moment is the '*Stena-Seacat HSS*'. She travels from Holyhead to Ireland and is capable of reaching speeds of up to 50 kt!!

Chapter 11
Ships of this Millennium

This chapter gives details mainly of Merchant ships delivered after December 1999.

Table 11.1 shows examples of the deadweight (dwt)/m³, main dimensions, speed and power for Gas Carriers, Ultra Large Crude Carriers (ULCCs), Passenger Liners and Container ships.

Table 11.2 shows examples of the dwt, Main Dimensions, speed and draft for RO-RO vessels, Very Large Crude Carriers (VLCCs), Bulk Carriers, General Cargo ships and Chemical Carriers.

Table 11.3 shows examples of VLCCs and ULCCs for vessels in existence together with ULCCs estimates up to a dwt of 1 000 000 tonnes. They indicate the tremendous size of these ships. Some of them are the *length* of five football pitches or six hockey pitches!!! Communication along the Upper Deck between members of the crew has to be by mobile, video-phone or unicycle. *Breadth Mld* (Br. Mld) can be similar to the length of a football pitch.

	dwt (tonnes)	LBP (m)	Br. Mld (m)	SLWL (m)	Service speed (kt)
Biggest Oil Tanker (*Jahre Viking* built in 1980)	564 739	440	68.80	24.61	13
Biggest RO-RO ferry (*Ulysses* built in 2001)	9665	192.4	31.20	6.40	22
Biggest Ore Carrier (*Peene Ore* built in 1997)	322 398	320	58	23.00	14.70
Biggest Container ship (*Hong Kong Express* built in 2002)	82 800	304	42.80	13.00	25.3
Biggest Passenger Liner (*QM2* built in 2003)	GT = 150 000	345	41.16	10.00	30
Fast Passenger ship (*Stena Explorer* built in 1996)	1500	107.5, (LOA = 125)	40.00	4.50	40

Table 11.1 Ships of this Millennium for Gas Carriers, ULCCs, Passenger Liners and Container ships

Delivery date	Name of vessel	dwt (tonnes)	LBP, L (m)	Br. Mld, B (m)	L/B value	Depth, D (m)	Draft, H (m)	H/D value	Speed (kt)	P_S or P_B (kW)
Gas Carriers										
Jun-00	Berge Danuta (LPG)	47 760	218.6	36.40	6.01	22.00	11.25	0.51	18.00	17 640
Jun-02	LNG Rivers	67 100	274.0	48.00	5.71	26.50	11.15	0.42	19.75	17 277
Jun-02	Stena Caribbean (LPG)	8 600	117.1	23.80	4.92	9.50	6.10	0.64	13.50	5 860
Nov-02	British Trader (LNG)	68 100	266.0	42.60	6.24	26.00	11.35	0.44	20.10	29 467
Jun-03	Berge Everette (LNG)	70 300	266.0	43.40	6.13	26.00	11.40	0.44	19.50	26 985
ULCCs										
Mar-02	Hellespont Alhambra	407 469	366.0	68.00	5.38	34.00	23.00	0.68	16.00	36 911
1980–2003	Jahre Viking	564 769	440.0	68.80	6.40	29.80	24.61	0.83	13.00	36 778
Passenger Liners										
Jun-00	Millennium	8 500	262.9	32.20	8.17	10.60	8.00	0.75	24.00	9 000
Oct-00	Olympic Champion	6 500	185.4	25.80	7.19	9.80	6.75	0.75	29.00	37 578
Jun-01	Silver Whisper	3 000	161.8	24.80	6.53	8.40	6.12	0.75	19.00	11 638
Feb-02	The World	5 058	173.0	29.80	5.81	9.20	6.90	0.75	17.00	8 235
Mar-02	La Superba	9 750	186.2	30.40	6.13	10.15	7.80	0.75	31.00	67 200
Dec-03	QM2	GT = 150 000	345.0	41.16	8.38	13.70	10.00	0.73	30.00	117 436
Container ships										
May-00	Sea-land New York	61 700	292.0	40.00	7.30	24.20	12.00	0.50	25.80	57 870
Sep-00	OOCL San Francisco	49 717	264.0	40.00	6.60	24.00	12.00	0.50	25.20	55 681
Sep-01	APL Venezuela	32 209	210.2	32.24	6.52	18.70	10.50	0.56	22.04	19 597
Oct-02	Hong Kong Express	82 800	304.0	42.80	7.10	20.23	13.00	0.64	25.30	69 646
Jul-03	Carmel Ecofresh	16 494	174.5	25.14	6.94	16.40	9.30	0.57	21.00	16 520

Source: Significant Ships of 2000, 2001, 2002 and 2003. Published annually by RINA, London.

Table 11.2 Ships of this Millennium for Roll-on/Roll-off vessels, VLCCs, Bulk Carriers, Cargo ships and Chemical Carriers

Delivery date	Name of vessel	dwt (tonnes)	LBP, L (m)	Br. Mld, B (m)	L/B value	Depth, D (m)	Draft, H (m)	H/D value	Speed (kt)	P_S or P_B (kW)
Roll-on/Roll-off vessels										
Aug-00	Crystal Ray	16 050	188.0	32.26	5.83	14.00	9.00	0.64	20.20	11 593
Apr-01	Pride of Rotterdam	9 268	203.7	31.50	6.47	9.40	6.05	0.64	22.00	28 199
Apr-02	Grand Pioneer	19 120	190.0	32.26	5.89	34.06	8.55	0.25	20.00	14 160
Apr-03	Norrona	5 230	152.4	30.00	5.08	9.10	6.30	0.69	20.80	21 600
May-03	Pasco Paoli	10 375	163.4	30.50	5.35	9.80	6.50	0.66	24.10	38 800
VLCCs										
Jan-00	Ubud	279 999	316.6	60.00	5.28	28.90	19.10	0.66	16.10	27 572
Jun-00	British Progress	301 440	320.0	58.00	5.52	31.25	22.30	0.71	15.00	25 849
Apr-01	Stena Victory	266 200	320.0	70.00	4.57	25.60	16.76	0.65	16.90	32 063
Oct-01	Harad	284 000	318.0	58.00	5.48	31.25	21.40	0.68	16.40	32 825
Jul-03	Capricorn Star	299 000	319.0	60.00	5.32	30.50	21.50	0.70	14.60	25 849
Bulk Carriers										
Jun-00	Jin Hui	44 579	182.0	32.26	5.64	16.69	10.75	0.64	14.80	8 206
Jun-01	Kohyohsan (Ax)	157 322	279.0	45.00	6.20	24.10	16.50	0.68	14.70	14 710
Jul-03	IVS Viscount	32 687	172.0	28.00	6.14	15.20	10.20	0.67	14.50	7 650
Jan-04	Tai Progress	64 000	217.0	32.26	6.73	19.50	12.20	0.63	14.50	9 996

(Continued)

Table 11.2 (Continued)

Delivery date	Name of vessel	dwt (tonnes)	LBP, L (m)	Br. Mld, B (m)	L/B value	Depth, D (m)	Draft, H (m)	H/D value	Speed (kt)	P_S or P_B (kW)
Cargo ships										
Nov-01	Salico Frigo (Reefer)	6 150	120.0	18.80	6.38	10.28	7.50	0.73	17.00	1 500
Apr-02	Arklow Rally	4 500	85.0	14.40	5.90	7.35	5.79	0.79	11.50	1 800
XXX	SD14	15 025	137.5	20.40	6.74	11.80	8.90	0.75	15.00	5670
XXX	SD20 (3rd generation)	19 684	152.5	22.80	6.69	12.70	9.20	0.72	15.00	6714
Chemical Carriers										
Feb-01	Falesia	4 500	91.5	15.40	5.94	7.80	6.00	0.77	14.00	3 060
Mar-01	Isola Blue	25 000	155.6	26.90	5.78	13.60	10.00	0.74	15.20	7385
Mar-02	FS Vanessa	15 500	134.0	23.00	5.83	15.70	8.30	0.53	14.00	6300
Jul-02	Tarantella	40 600	176.0	32.20	5.46	17.20	10.98	0.64	15.02	7680
Mar-03	Cosmo	5 884	95.0	16.98	5.59	8.60	6.60	0.77	14.00	3 360

Source: Significant Ships of 2000, 2001, 2002 and 2003. Published annually by RINA, London.
XXX as per Standard General Cargo ships (Fairplay).

Table 11.3 Giants of the sea

Oil Tanker dwt (tonnes)	LBP, L (m)	Br. Mld, B (m)	L/B value	Draft Mld (m)	C_B value	C_D value	Speed (kt)	P_S or P_B (kW)	No. of propellers (rpm)
As built and proposed									
200 000	314	50.60	6.21	17.40	0.838	0.843	16	20 890	1@80
250 000	328	53.40	6.14	19.50	0.842	0.847	16	24 620	1@80
250 000	328	53.40	6.14	19.50	0.842	0.847	16	33 200	2@70
300 000	340	55.80	6.09	21.40	0.847	0.852	16	39 540	2@70
350 000	353	58.40	6.04	22.70	0.850	0.855	16	44 330	2@70
400 000	366	61.00	6.00	23.80	0.854	0.859	16	52 220	2@70
500 000	392	66.10	5.93	25.20	0.862	0.867	16	63 410	2@70
Via proposals only									
500 000	392	66.10	5.93	25.20	0.862	0.867	16	48 120	3@60
600 000	416	71.10	5.85	26.10	0.868	0.873	16	52 970	3@60
700 000	435	76.50	5.68	26.80	0.872	0.878	16	57 440	3@60
800 000	454	81.70	5.56	27.30	0.875	0.881	16	61 920	3@60
900 000	472	86.40	5.46	27.90	0.876	0.881	16	66 020	3@60
Megatonne design	492	90.10	5.46	28.20	0.876	0.881	16	70 120	3@60

For dwt up to 500 000 tonnes, L = 262 + 0.26 (dwt/1000) m and B = (L/5) − 12.2 m.
The above values were obtained from curves plotted by the author, from vessels (as built) and proposed designs worldwide.

Standard ships

Over the last 40 years many ships have been designed and built as Standard ships (S/S). This means that several ships have been built from the same design, drawings and specification.

Each S/S will have the same Main Dimensions, the same hull form, the same power, the same service speed and a similar layout of accommodation and navigation spaces. They are similar to buying an 'off the peg' suit, instead of buying a 'made to measure' suit.

There is now several catalogues for S/S. One can scan through them and see examples such as the SD14, the SD20, the *Unity*, the *Cartago*, the STATT 55, the B26, the MP25, the *Clyde*, the *Freedom*, the StaFF20, the *Pioneer*, the Hull 741 etc. Because of continuing improvements in ship technology over the years, some of these designs are now third or fourth generation of the original design.

The advantages of S/S can be summarised as follows:

1. Ship dwt and service speed has already been previously attained.
2. Ship-model tests have been completed and paid for.
3. The steel order book for the steel work has already been finalised.
4. The shipbuilding Network Analysis programme is already in place.
5. All calculations and drawing office work has been already completed.
6. All computer-cutting tapes for prefabrication work in the shipyard are stored ready for further use.
7. Assembly problems have been met and answered for previously built S/S.
8. All the design work has been completed and approved by Classification Surveyors.
9. A shipowner can view a previously built S/S to observe and appraise what is being supplied.
10. A shipowner can request changes to the S/S. This may be the fitting a Diesel Engine instead of a Steam Turbine installation or changes in the layout of the accommodation. When this occurs, negotiation takes place for revision of cost and Delivery Date. All agreed modifications are entered in the Shipbuilder's Ship Specification book.
11. Ship Trials date and Delivery Date can be given with increased certainty.
12. Based on building experience repeatedly gained on previous S/S, there may well be a quicker delivery time and a higher standard of workmanship.
13. Like a car purchaser, a shipowner can seek opinions from a shipowner already operating an S/S.
14. Similar to a standard car, there is a much greater chance of obtaining spares.
15. An *appreciable lower first cost*, compared to that for a 'one-off' ship. It is just like purchasing an 'off the peg' suit.

A note of caution: There are disadvantages with the building of S/S. Because of repetition of work monotony can develop. Shipbuilders in the yard can become blasé because they have done it so many times. It occurs on car assembly lines.

Because of this, mistakes can occur causing hold ups in the prefabrication assembly line. Time costs money. The quality control men within the ship-yard must therefore be extra vigilant.

One shipyard successfully answered these problems by switching their workers from being on an SD14 assembly line to being on a B26 assembly line and vice versa, at say 6-month intervals.

Chapter 12

Ship Trials: a typical 'Diary of Events'

This chapter is really an introduction to the next four chapters that deal with measuring ship performance. They are:

- Speed performance on the measured mile.
- Endurance and fuel consumption tests.
- Manoeuvring tests and stopping characteristics.
- Residual tests.

In this Diary of Events, consideration has been given to covering the pro-forma sheet, to the completion of the vessel construction at the shipyard, to the dry-docking of the ship prior to the actual trials, to the final tests, to the supplies to the ship and to the 3-day actual Trial Trip programme itself.

For the shipbuilder 'Ship Trials' are financially very important. When successfully completed, it means that an agreed percentage (15–20%) of the first cost of the new ship can be handed over by the shipowner.

During the building of a ship, the work is constantly monitored by ship surveyors. In the UK, Lloyds Surveyors generally examine the *strength* of the ship. This entails scantling plans inspection, types of material, types of welded joints, water-tightness tests, freeboard markings, electrical systems, fire prevention detection and extinction requirements, pumping and piping systems, boilers and other pressure vessels, refrigerated cargo installations, annual and special surveys, carriage of engineering spares, machinery and propulsion units, etc.

The Department of Transport (DfT) Surveyors generally look after the *safety* aspects. This entails examining all life-saving appliances, lifeboats, life-rafts, accommodation, heating, lighting and ventilation arrangements, gross and net tonnage values, carriage of grain arrangements, fire prevention detection and extinguishing arrangements, navigation lights, cargo ventilation systems, sound signals and freeboard markings, etc.

Plans have to be submitted for approval to both Classification Societies. Surveyors have offices within each shipyard to ensure that each ship is 'as built' according to these approved plans. If not, then approval is not given.

In recent years there has been a blurring and sharing of the responsibilities of Surveyors, so much so that the blanket term 'ship surveyor' is more commonly used. Whatever their title is, their job is quality control.

Pro-forma details

For the trials, a pro-forma sheet has to be filled in. This sheet gives details on:

- Ship's name, type of ship, service speed, service power in K_W and number of screws.
- General data, such as names of shipbuilder and shipowner, General Particulars of the ship, date and location of the Ship's Trial.
- Brief statement of the loaded condition of the ship before and after the Ship Trials.
- Power details relating to the propeller shaft.
- Propeller and stern arrangements, relating to diameter of propeller(s), number of blades, type of propeller pitch, type of stern and type of rudder fitted.

Completion of vessel construction programme

- Commence preliminary inspection of accommodation spaces (DfT).
- Oil Fuel tanks complete and closed.
- Oil Fuel, air and overflow, filing and contents gauge systems completed.
- Test boat winches (portable motor).
- Ship diesel oil.
- Boat lowering test and motorboat trial (DfT).
- Preliminary test of windlass and windlass.
- Ship diesel alternator lubricating oil.
- Basin trial of anchors and cables.
- Preliminary run of diesel alternators.
- Basin trial of steering.
- Diesel alternator electrical trials.
- Emergency diesel alternator trials.
- Commence ballasting for dry-docking.
- Preliminary tests for mechanical ventilation system (DfT).
- Check power on deck.
- Inspection of navigation lights (DfT).
- Check life-saving equipment, boat equipment (including onboard stores) (DfT).
- Rig and stow accommodation ladder.
- Test derricks.
- Inclining experiment: add or deduct final group weights to complete ship's final lightweight (DfT Surveyor in attendance). This experiment is also known as the 'stability test' in some shipyards.
- Before Sea Trials, whilst the ship is still in the basin to get a high load, Surveyors require an electrical overload test plus reverse trips. The overload is sometimes achieved by controlled immersion of a steel plate into a water tank.

Dry-docking of vessel

- Power required on deck.
- Inspect bottom shell and sideshell.
- Clean, touch up and coat shell plating with paint.
- Clean and inspect propeller.
- Deratisation certificate inspection.
- Check freeboard markings. (Lloyds and DfT Surveyors both involved with this.)
- In the case of a controllable-pitch propeller having been fitted, whilst the ship is still in dry-dock, the surveyor has to confirm that that the pitch indication is the same reading on the bridge, at the Oil Distribution box and at the propeller. In other words, the propeller pitch readings have been correctly calibrated at these three locations.

Undock to basin

- Power required on deck.
- Test of galley and pantry gear.
- Test Fire and Washdeck line service and emergency fire pump.
- Cabin heating test (DfT).
- Test hot water and sanitary water systems (DfT).
- Bilge pumping test (Lloyds).
- Fill freshwater tanks.
- Cooling down test of domestic refrigeration chambers.
- Basin Trial of main machinery, test whistles, telegraphs, etc.
- Commence ballasting and fuelling for Ship Trials.
- Test domestic refrigerators and drinking fountains.
- Inspection of fire-fighting equipment, test wireless transmitter, direction finder radar, echo-sounder equipment, etc. (DfT).
- Test fire alarms and emergency valve closing arrangements.
- Test electrical and emergency lighting, etc. (DfT).
- Check tonnage and crew space markings (DfT).
- Check compasses and navigational instruments (DfT).
- Telephone communication system complete and tested (DfT).
- All accommodation spaces complete and inspected (DfT).
- Cool down refrigerated chambers for Ship Trials stores.
- Ship Trials personnel 'sign-on' articles.
- Take on stores and complete arrangements for Ship Trials.
- Transfer of water in cargo tanks completed.
- Vessel leaves basin for shipbuilder's Ship Trials.

Ship Trial programme

Day 1
- Leave basin.
- Commence ballasting at Crosby, near to Liverpool (if required).

- Arrive at bar. Adjust compasses, direction finder, echo-sounder and sonar.
- Start and complete Anchor Trials.
- Proceed to Arran Measured Mile and during passage, commission machinery controls.

Day 2

- Arrive at Arran and carry out four double runs on the measured mile.
- Vessel to be photographed during the measured mile runs.
- On completion, increase to service power and carry out the following trials in the Firth of Clyde:
  - Steering Trials.
  - Turning circle diameter (TCD) trials and Zig-zag manoeuvres.
  - Crash-stop manoeuvres and Astern Trials.
  - Return to the River Mersey, carry out 6-h Endurance Trials on the way.
  - On completion of Endurance Trials, de-ballast as required.

Day 3

- Arrive at the Mersey Bar.
- Proceed up river and anchor off shipyard.
- Tender in attendance to take Ship Trial party ashore.
- Enter the wet Basin.

Certificate of Registry and the Carving Note

After the ship has completed the trials, she is ready to be formally handed over to the shipowner. A Certificate of Registry is prepared by the DfT Surveyor. It is then forwarded to the Registrar at the intended Port of Registry. This certificate gives all the particulars and Main Dimensions of the new ship.

Before the actual delivery if this Certificate of Registry to the shipowners, the Registrar issues a Carving Note. This gives details of markings that should be on the ship, namely:

- Ship's name.
- Port of Registration.
- Official ship number.
- Net tonnage value (to be marked on a main beam).

After satisfactory marking of these markings are certified by the DfT Surveyor, the Certificate of Registry can then be supplied by the Registrar. All that remains now is an agreed premium (10–15% of the first cost of the ship), to be paid by the shipowner to the shipbuilder.

Chapter 13

Ship Trials: speed performance on the measured mile

When a vessel is completed, various trials take place in order to confirm that the ship's performance is as specified by the shipowner when the Memorandum of Agreement was signed. These trials are also known as Acceptance Trials.

Part of these trials are Speed Trials over the measured mile. This is one nautical mile of 1852 m. This is undertaken to verify that the new vessel can attain a certain speed for a given shaft or brake power.

Due to the displacement being lower than the fully loaded displacement, the Trial speed is about 0.75–1.00 kt greater than the specified designed service speed. See later worked example to illustrate this.

Precautions

In order to obtain reliable results during Ship Trials, the following precautions should be observed:

1. The vessel should be dry-docked immediately before the trials. The hull should be cleaned and painted. The number of days out of dry-dock prior to the trials should be noted.
2. Trials should be carried out in calm weather, with little or no wind. The presence of wind requires a correction factor that is difficult to estimate.
3. Trials should be carried out in water where the depth of water will not influence the ship's resistance. A ship is said to be in deep water when the depth of water is greater than the depth of influence (F_D). Table 18.1 and Figure 18.4 in Chapter 18 show depths of influence (F_D) for several types of Merchant ships.
4. Because the trial displacement is less than the fully loaded displacement, care must be taken to ensure that the propeller tips are fully immersed. If not, 'racing' will occur.
5. Sufficient length of run should be given before the ship comes onto the measured mile (see Figure 13.1).
6. Distance from the shore should be kept as constant as possible. This is to minimise tidal differences. The speed of the ship is relative to the land. It

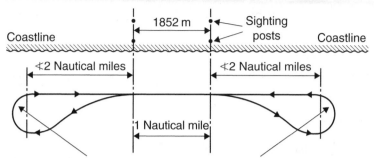

Fig. 13.1   Manoeuvres on the measured mile.

can be determined by using sighting posts with the aid of stopwatches (see Figure 13.2).

For example, if it takes 3 min to travel over the measured mile, then the ship's speed is 60/3, i.e. 20 kt. At least three independent observers should record the time of transit on the mile.

In recent years, large Bulk Carriers, Very Large Crude Carriers (VLCCs) and Ultra Large Crude Carriers (ULCCs) have had to seek deeper waters for there trials due to them having very large draft values. If shore sighting posts are not used, then one of the following methods may be used:

- Take sightings from the ship onto a fixed buoy.
- Use electronic 'Decca fix' procedure.
- Use Global Positioning System (GPS) method.

Radiometric distance measuring systems such as the Tellurometer, the Decca Navigator, Hi Fix and others can provide an alternative method of measuring ship speed to the required accuracy of +0.10 to −0.10 kt, without incurring many of the problems associated with measured mile courses. These systems have been shown to give close agreement with stopwatched times.

Radiometric systems may be used at suitable locations off the coast. They are not affected by adverse weather conditions such as fog or sleet. The operating range of these devices enable Ship Trials to be conducted at greater distances offshore in deeper water, without approach runs and accompanying manoeuvring problems.

7. An accurate straight course should be kept. Rudder helm movement should be 5°P to 5°S over the measured mile.

8. Several runs should be made for these Speed Trials (say 4–8) over the measured mile. This is so comparisons and conclusions for the ship speed can be made.

9. The Ship's Trial draft should correspond with the ship-model's draft in trial condition of loading tested and run in a towing tank or in a flume. This is in order to obtain a geosim relationship comparison.

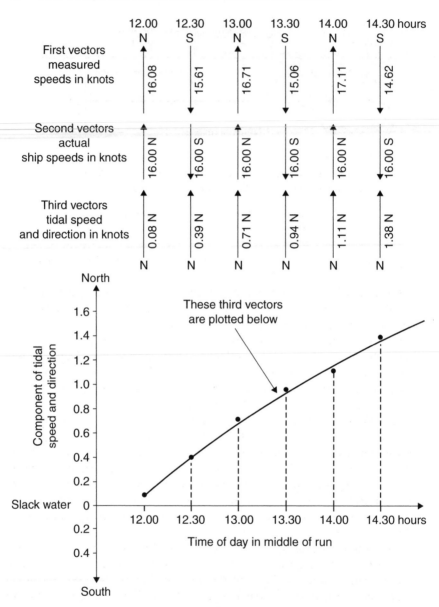

Fig. 13.2   Tidal component of speed and direction.

Data measured

During these runs on the measured mile, as certain amount of data is measured and noted, namely:

(a) Speed of ship over the ground.
(b) Propeller and engine rpm, corresponding to these speeds.
(c) Shaft power or brake power, corresponding to the speeds.

(d) Wind force and direction. Tide speed.
(e) Condition of the sea, water depth and water density.
(f) Log of the ship.
(g) Distance from the coastline.
(h) Vessel's drafts: aft, amidships and forward.
(i) Thrust at the thrust block.
(j) Apparent slip (discussed in Chapter 6).

Estimation of a Ship's Trial speed

Method 1

The ship's speed on the measured mile will contain a component of the tidal and current effects. This component will change with time and so will be different for each trial run.

Some method must be used to separate these effects and so obtain the true speed of the ship on each run.

If the runs are made with a common interval of time between them, say 30 min, then a mean of means calculation will give a true ship speed.

Consider four runs on the measured mile spaced 30 min apart. Assume that the measured speeds are V_1, V_2, V_3 and V_4 kt.

$$\text{Mean of means speed} = \text{True speed of ship} = \frac{V_1 + 3V_2 + 3V_3 + V_4}{8} \text{ kt}$$

The figure of eight represents the summation of Dr Simpsons's multipliers of $1 + 3 + 3 + 1$.

If six runs are made then,

$$\text{Mean of means speed} = \text{True speed of ship}$$
$$= \frac{V_1 + 5V_2 + 10V_3 + 10V_4 + 5V_5 + V_6}{32} \text{ kt}$$

The figure of 32 represents Dr Simpsons's multipliers of $1 + 5 + 10 + 10 + 5 + 5 + 1$.

Worked example 13.1 shows this method. Note the extraction of the Tidal speed and direction component, mathematically and graphically. It has to be a component, unless the direction of the ship's run is right in line directionally with the movement of the tide.

Worked example 13.1

A new ship on Acceptance Trials makes six runs as shown below at 30 min intervals of time, beginning at 12.00 noon.

(a) Calculate the average speed and the true speed for this ship.
(b) Draw a graph of tide's speed and direction component against time of day in middle of run.

Time of day in middle of run	12.00	12.30	13.00	13.30	14.00	14.30
Direction of each run	North	South	North	South	North	South
Measured ship speed, V (kt)	16.08	15.61	16.71	15.06	17.11	14.67
Tidal speed and direction	0.08 N	0.39 N	0.71 N	0.94 N	1.11 N	1.38 N

$$\text{For six runs, average speed} = \frac{16.08 + 15.61 + 16.71 + 15.06 + 17.11 + 14.67}{6}$$

$$= 15.865 \text{ kt}$$

This 15.865 kt incorrectly assumes that a tide's movement to be a series of straight lines instead of being what it is, a parabolic plot. True speed formula does assume that the tidal effect is parabolic, as of course it is in real life.

$$\text{True speed of ship} = \frac{V_1 + 5V_2 + 10V_3 + 10V_4 + 5V_5 + V_6}{32} \text{ kt}$$

$$\text{Speed} = \frac{(1 \times 16.08) + (5 \times 15.61) + (10 \times 16.71) + (10 \times 15.06) + (5 \times 17.11) + (1 \times 14.67)}{32}$$

So the Ship's Trial speed = 16.00 kt   (it now contains no tidal or current effects).

With Method 1, the time interval between each run must be similar. In practice, it is generally very difficult to time the runs with a constant interval of time. Consequently another method or technique for evaluating the true speed on each ship run must be used.

Method 2

Worked example 13.2

The performance data for a Ship's Trial on the measured mile for four runs is shown in Table 13.1.

A graph of revolutions per nautical mile (RPNM) against time of day is drawn. This is shown in Figure 13.3. One line is for the ship when travelling North, that is on runs 1 and 3. The other line is for when the ship is travelling South, that is on runs 2 and 4.

Figure 13.3 illustrates how these two lines are bisected to obtain a third line. This is called the mean RPNM line. It is denoted the 'Nm' line. Lift off the value for each run. They are 415.0, 418.5, 422.5 and 426.2, as shown in column 7 of Table 13.1.

$$\text{Ship's true speed on each run} = \frac{N \times 60}{Nm} \text{ kt}   \text{See column 8 in Table 13.1}$$

Table 13.1   Table of Ship Trial measurements and results

Run and direction	Time of day in middle of run	Obs. time on measured mile T (min)	Ship speed V (kt)	Propeller rpm (N)	RPNM (N × T)	Mean RPNM (Nm)	True speed of ship (kt)	Tidal speed and direction (kt)
1 N	13.01	5.05	11.90	86.10	435	415.0	12.45	0.55 S
2 S	13.27	4.68	12.81	86.88	406	418.5	12.46	0.35 S
3 N	14.02	4.97	12.07	85.85	426	422.5	12.19	0.12 S
4 S	14.29	4.30	13.90	100.15	431	426.2	14.10	0.20 N
*	*	*	+	*	+	+	+	+

*Denotes given information from recorded data on trials.
+ Denotes values obtained from the given information.

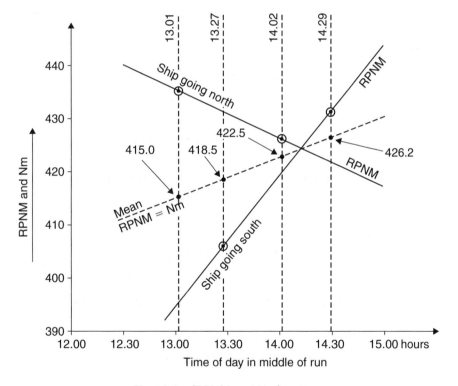

Fig. 13.3   (RPNM and Nm) ∝ time.

For the four runs, the ship speeds were shown to be 12.45, 12.46, 12.19 and 14.10 kt.

The component for the tide's speed and direction can be extracted previously demonstrated in an earlier example. This is shown in column 9 of

Table 13.1 and also shown graphically in Figure 13.4. Note in this Figure 13.4 that slack water occurred at 14.09 h. Slack water occurs when the speed of the tide is zero.

If the Ship Trials are held over a long period of time it is possible to draw the tidal speed and direction as depicted in Figure 13.5. In certain parts of the world, slack water will constitute part of the tidal curve where it is zero over a definite period of time. This is shown in Figure 13.5(b).

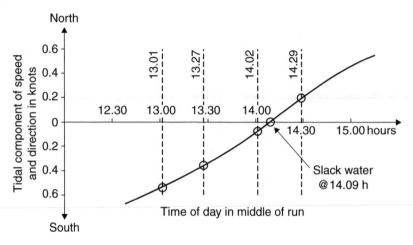

Fig. 13.4   Tidal component of speed and direction ∝ time.

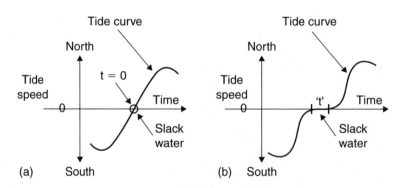

Fig. 13.5   Two graphs depicting slack water.

Worked example 13.3

A General Cargo vessel is 14 500 tonnes displacement when fully loaded, with a service speed (V_1) of 16 kt. On her trials she is run at full shaft power but only loaded up to 76% of her fully loaded displacement. Calculate the Trial speed (V_2). Assume that the Admiralty Coefficient (A_C) is similar for both conditions of loading.

$$A_C = \frac{W^{2/3} \times V^3}{P_S} \quad \text{for both conditions}$$

Also P_S is same for both conditions and so cancel out on both sides.

So
$$14\,500^{2/3} \times 16^3 = (14\,500 \times 76\%)^{2/3} \times V_2^3$$

Hence
$$V_2^3 = \frac{594.6 \times 4096}{495.2} \qquad V_2 = 4918^{1/3}$$

So
$$\text{Trial speed} = 17\,\text{kt}$$

Note that the speed expected on trials will be 17 kt, i.e. 1 kt greater than the ship's service speed of 16 kt. The 17 kt speed should agree with that predicted from ship model experiments simulating the 76% loading of the ship.

It should also be noted that the Trial speed on the measured mile is not always attained. One example, experienced by the Author, resulted in a ship having to go on a second round of Speed Trials.

The main reason was a long delay between dry-docking the ship prior to the trials and actually going on the trials. Freak weather caused heavy siltation on the hull of the vessel. This caused extra frictional resistance and caused the ship to under perform.

A second dry-docking and clean up of the hull was arranged at the ship-builder's expense. Minimum delay was made before taking the vessel onto the measured mile. The vessel exceeded her expected and predicted Trial speed. The shipowners were happy. So were the shipbuilders!!

Progressive Speed Trials

In these trials, the ship is run at 100% MCR, 85% MCR, 75% MCR and at 50% MCR. During these runs, ship speed and engine rpm are measured and recorded.

MCR stands for Maximum Continuous Rating of the ship's engine. It is about 85% of the maximum possible power. It is at MCR power that the service speed is designed for. The extra 15% power is in effect a reserve of power, to cover for life-threatening emergencies, or loss of speed due to wind and waves, etc.

MCR power is the power generated hour after hour, day after day or week after week, as required. In other words, 'a continuous power.'

Figure 13.6 shows the extra power required for a 2 kt increase in speed from say 12 to 14 kt and from 14 to 16 kt. *Twice* the extra power is needed for the same increment of speed at the higher ship speeds. This means a much greater extra cost for the machinery and increased oil fuel consumption per day. They are definitely points to seriously consider, when deciding upon a higher than usual service speed, for example with new Container vessels and new Passenger Liners.

Table 13.2 gives particulars for progressive speed trials for a twin-screw ship, where the measured speeds ranged from 10.87 up to 18.03 kt. Mean values were evaluated for ship speed, shaft power, thrust and propeller revolutions. Figures 13.7–13.9 show curves produced by the author of this book to graphically illustrate relationships between these values.

Figure 13.7 shows the graph of shaft power against thrust. Figure 13.8 shows shaft power against ship speed. Figure 13.9 shows thrust against propeller revolutions.

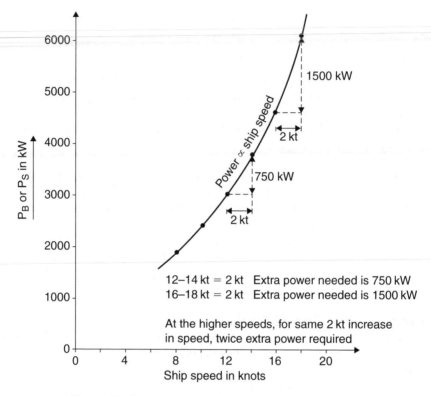

Fig. 13.6   Progressive Speed Trials. Power ∝ speed.

Table 13.2   Progressive Speed Trial results for a twin-screw ship

Run and direction	Ship speed V (kt)	Shaft power P_S (kW)	Thrust T (kN)	Propeller shaft N (rpm)	N^2	N^3
1 N	10.87	2916	492.22	66.76		
Mean values	11.22	2979	503.68	66.94	4481	300 000
2 S	11.56	3042	515.14	67.11		
3 N	13.82	5928	786.16	84.20		
Mean values	13.87	5943	786.16	83.79	7021	588 300
4 S	13.92	5958	786.16	83.37		
5 N	15.78	8926	1018.32	95.45		

(Continued)

Table 13.2   (*Continued*)

Run and direction	Ship speed V (kt)	Shaft power P_S (kW)	Thrust T (kN)	Propeller shaft N (rpm)	N^2	N^3
Mean values	15.63	8650	1003.88	94.30	8892	838 600
6 S	15.47	8374	989.43	93.15		
7 N	17.48	12 819	1313.26	106.38		
Mean values	17.15	12 860	1320.73	106.35	11 310	1 202 900
8 S	16.81	12 901	1328.20	106.31		
9 N	18.03	13 989	1999.94	109.18		
Mean values	17.33	14 001	1406.42	109.06	11 894	1 297 200
10 S	16.62	14 012	1412.90	108.94		

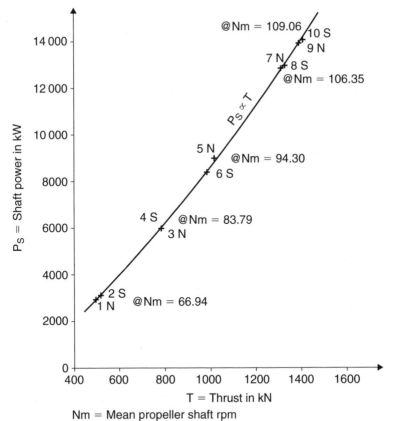

Nm = Mean propeller shaft rpm
10 Runs made: 5 North and 5 South
4S denotes 4th run in South direction

Fig. 13.7   Shaft power ∝ thrust for a twin-screw ship.

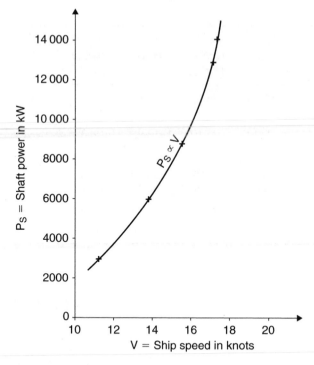

Fig. 13.8   Shaft power ∝ speed of ship for a twin-screw ship.

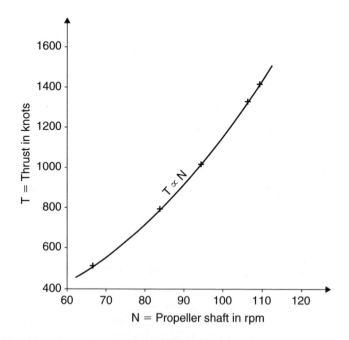

Fig. 13.9   Thrust ∝ propeller revolutions for a twin-screw ship.

Questions

1 List the precautions to be taken before and during Ship Trials for a new vessel.

2 When considering the runs on the measured mile, list the data that is measured and noted during Ship Trials.

3 On Acceptance Trials, a Bulk Carrier makes six runs as shown in the table below. The runs are made every 45 min apart.

Run and direction	1 N	2 S	3 N	4 S	5 N	6 S
Measured speed over the ground (kt)	17.08	16.61	17.71	16.06	18.11	15.62

Calculate:

(a) True speed of ship on each run.
(b) Tidal component's speed and direction for each run.

4 On Ship Trials, a vessel makes six runs and records the following data:

Run and direction	Time of day in middle of run	Measured speed over the ground (kt)	Propeller (rpm)
1 N	12.30	12.52	101
2 S	13.00	11.70	101
3 N	14.30	14.30	127
4 S	15.00	13.96	127
5 N	16.15	13.91	133
6 S	16.45	14.09	133

Calculate the true speed of this ship for each run.

Chapter 14

Ship Trials: endurance and fuel consumption

In endurance and Fuel Consumption trials, the vessel is run at Maximum Continuous Rating (MCR) power for a fixed duration, say 6–24 h. During this period of time, the following information is measured and recorded:

- Fuel consumption in kg/kW hour.
- Propeller and Engine rpm.
- Indicated power (P_I) within the Engine Room.
- Feed water used.
- Engine oil pressures and temperatures.
- Gearbox oil pressures and temperatures.
- Engine water temperatures.
- Fuel water temperatures.
- Auxiliary units, to verify no presence of overheating.
- Monitoring of electrical gear units performance to required standards.

Engine settings during Ship Trials

The following matters are the responsibility of the Engine Room staff:

(a) On making a group of runs at a given speed, the original engine settings used when first approaching the measured distance should be rigorously maintained throughout the group. No adjustments should be made when turning about for the return runs.

(b) When altering propeller shaft revolutions or propeller pitch for a new group, the new settings should be made as soon as possible after completing the readings and certainly before finally turning to approach the measured mile itself. Again, they should remain unaltered throughout the group.

(c) *Engine-control settings*: In current practice, automatic speed governors ensure constant rpm during each group of runs. Where

governors are not fitted, the following precautions should be taken:
- *Diesel machinery*: The fuel-control setting should be left unaltered throughout the runs of a particular group.
- *Steam Turbine machinery*: It is essential that the manoeuvring valve settings should be left unaltered during the runs of a particular group. Every endeavour should be made to maintain steady boiler conditions.
  (d) When a controllable-pitch propeller is fitted, the pitch settings used when first approaching the measured mile should be left unaltered throughout the group of runs.

As well as testing machinery, the performance and endurance of Ship Officers may also be evaluated. This is ergonomics, the relationship between personnel and machines over a period of time. Working conditions and fatigue limits are often analysed. Often, valuable feedback of information is obtained and used in future policy-making decisions.

Fuel consumption values

Method 1
At the beginning of this Millennium, for Merchant ships, the fuel consumption values are of the order of:

$0.200 \, \text{kg/kWh}$ or $0.00480 \times P_S$ tonnes/day   for Steam Turbine machinery

$0.180 \, \text{kg/kWh}$ or $0.00432 \times P_B$ tonnes/day   for Diesel machinery installation

Method 2
To help approximate the fuel consumption per day (fuel cons/day) in tonnes/day, a fuel coefficient may be used. It is F_C, where:

$$F_C = \frac{W^{2/3} \times V^3}{\text{Fuel cons/day}} \quad \text{or} \quad \text{Fuel cons/day} = \frac{W^{2/3} \times V^3}{F_C} \text{ tonnes}$$

where:

W = ship's displacement in tonnes,
V = ship's speed in knots,
F_C = fuel coefficient, dependent upon type of machinery installed in the ship.

For Steam Turbine machinery: $F_C = 110\,000$ approximately

For Diesel machinery installation: $F_C = 120\,000$ approximately

Worked example 14.1

A ship's displacement is 14500 tonnes, speed 16 kt and shaft power (P_S) 5025 kW. Estimate the fuel cons/day given that 1000 kg is 1 tonne.

Method 1

Shaft power indicates the fitting of Steam Turbine machinery. Use 0.200 kg/kW h:

$$\text{So} \quad \text{Fuel cons/day} = \frac{0.200 \times 5025 \times 24}{1000} = 24 \text{ tonnes approximately}$$

Method 2

$$\text{Fuel cons/day} = \frac{W^{2/3} \times V^3}{F_C} \text{ tonnes} \qquad \text{Use 110 00 for } F_C$$

$$= \frac{14\,500^{2/3} \times 16^3}{110\,000} = 594.6 \times \frac{4096}{110\,000}$$

$$= 22 \text{ tonnes approximately}$$
$$\text{Similar to previous answer}$$

Worked example 14.2

For a Very Large Crude Carrier (VLCC), the displacement is 235000 tonnes, speed 15 kt and the P_S is 24750 kW. Estimate the fuel cons/day by two methods.

Method 1

Shaft power indicates the fitting of Steam Turbine machinery. Use 0.200 kg/kW h:

$$\text{So} \quad \text{Fuel cons/day} = \frac{0.200 \times 24\,750 \times 24}{1000} = 119 \text{ tonnes approximately}$$

Method 2

$$\text{Fuel cons/day} = \frac{W^{2/3} \times V^3}{F_C} \text{ tonnes} \qquad \text{Use 110 00 for } F_C$$

$$= \frac{235\,000^{2/3} \times 15^3}{110\,000} = 3808 \times \frac{3375}{110\,000}$$

$$= 117 \text{ tonnes approximately}$$
$$\text{Similar to previous answer}$$

Worked example 14.3

For the VLCC in the previous question, the Steam Turbine machinery was by a retrofit, fitted with a Diesel machinery installation having 24750 kW brake power (P_B). Estimate by two methods the new fuel cons/day with this retrofit. Assume again that the displacement is 235000 tonnes and the speed is 15 kt.

Method 1
Brake power indicates the fitting of Diesel machinery. Use 0.180 kg/kW h:

So   Fuel cons/day = $\dfrac{0.180 \times 24\,750 \times 24}{1000}$ = 108 tonnes   approximately

Method 2

$$\text{Fuel cons/day} = \frac{W^{2/3} \times V^3}{F_C} \text{ tonnes} \qquad \text{Use 120 00 for } F_C$$

$$= \frac{235\,000^{2/3} \times 15^3}{120\,000} = 3808 \times \frac{3375}{120\,000}$$

$$= 107 \text{ tonnes approximately}$$
$$\text{Similar to previous answer}$$

Information collected for Oil Tankers ranging from 30 000 to 260 000 tonnes dwt is as shown in Table 14.1.

Table 14.1   Fuel cons/day for a range of Oil Tankers (see Figure 14.1 for graphical representation of these tabulated values)

Deadweight (tonnes)	Power at the thrust block P_S or P_B (kW)	Steam Turbines (fuel cons/day) (tonnes)	Diesel machinery (fuel cons/day) (tonnes)
30 000	6 933	33	30
40 000	9 139	44	39
50 000	10 646	51	46
60 000	12 187	58	52
80 000	15 077	72	65
100 000	16 793	81	73
120 000	18 856	91	81
140 000	20 129	97	87
160 000	21 986	106	95
180 000	23 144	111	100
200 000	24 224	116	105
220 000	25 241	121	109
240 000	26 156	126	113
260 000	26 003	125	112

Conclusions

By fitting Diesel machinery in a ship of similar power, displacement and speed, a saving of about 10% in the daily fuel consumption can be achieved. Of course the differences in the cost of fuel/tonne must be taken into account plus the size of the machinery arrangement installed in the ship.

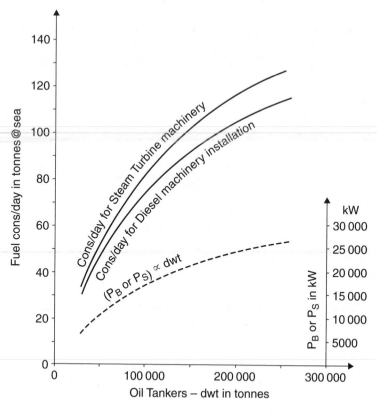

Fig. 14.1   Fuel consumption and power for Oil Tankers.

For General Cargo ships the fuel consumption is about 20–25 tonnes/day. For the Sealand Ltd Container ships operating at service speeds of 26 kt, it was up to 600 tonnes/day. For the Queen Mary 2 sailing on her maiden voyage in January 2004 from Southampton to Fort Lauderdale with a service speed of 30 kt, the fuel consumption was up to 800 tonnes of diesel fuel/day.

Questions

1 What information is measured and noted on endurance and Fuel Consumption trials?

2 Three Oil Tankers are fitted with Diesel machinery. Their fully loaded deadweights are 85 000, 150 000 and 225 000 tonnes. Estimate the approximate fuel cons/day for each tanker.

3 What are the fuel cons/day in terms of the power (P_S or P_B) at the thrust block for a ship with Steam Turbine machinery and a ship with Diesel machinery?

4 Estimate the power at the thrust block for three Oil Tankers having a fully loaded dwt of 75 000, 135 000 and 250 000 tonnes.

Chapter 15

Ship Trials: manoeuvring trials and stopping characteristics

This chapter will cover the following parts of the Ship Trials:

1. Spiral manoeuvre.
2. Zig-zag manoeuvre.
3. Turning circle diameter (TCD) trials.
4. Crash-stop manoeuvres.

Spiral manoeuvre

This work was first presented to the Institute of Naval Construction in Paris by Prof. J. Dieudonné. He started the manoeuvre by having a steady rudder helm of say 15° to Starboard. This was held until the ship's rate of change of heading became steady. The rudder helm was then reduced to 10° to Starboard, which again was held until the ship's steady rate of heading was reached (see Figure 15.1).

This procedure was repeated, passing through zero, onto 15° to Port, before returning to a rudder helm of 15° to Starboard. These points were then plotted. They produced a hysteresis loop within which directionally unstable conditions apply. The propeller rpm are held constant throughout this spiral manoeuvre.

If rudder helm is moved within O–T to Starboard and O–U to Port, the ship will respond capriciously, changing her heading to either side indiscriminately. This makes course keeping very difficult to achieve. The reason for this is due to hydrodynamic imbalance.

For some ships, the distance U–T in Figure 15.1 is zero. If so, the ship is directionally stable.

Zig-zag manoeuvre

These manoeuvres are conducted to provide a measure of a ship's response to rudder movement. The larger the rudder, the quicker will be the response.

Starting with the ship's heading on a straight course, the rudder is moved to say 20° to Port. It is held until the ship's heading has responded and moved also to 20° to Port. When this happens, the rudder is moved to 20° to Starboard. It is held until the ship's heading has responded and moved this time to 20° to Starboard. It will be found that the ship will have an overshoot value of 8° to 10° (see Figure 15.2). In shallow waters, this overshoot value will be greater.

Provided the vessel is directionally stable, the smaller the overshoot value and the shorter the time interval between successive rudder orders, the more efficient is the response of the ship. This is called 'controllability.'

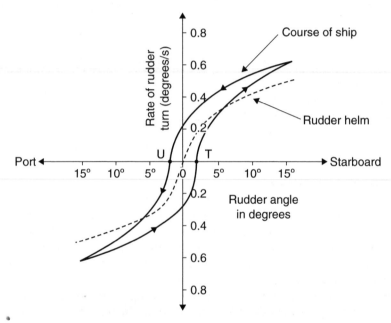

Fig. 15.1   Dieudonné spiral manoeuvre. Propeller revolutions constant throughout.

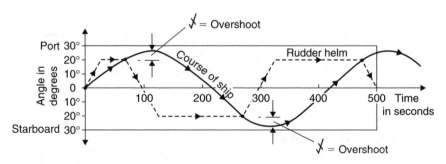

Fig. 15.2   Zig-zag manoeuvres. Ship run at full speed. (•) denotes change of rudder helm.

Turning circle diameter trials

This manoeuvre is carried out with the ship at full speed and rudder helm set at 35°. The ship is turned completely through 360° with say Starboard rudder helm and then with Port rudder helm (see Figure 15.3). There will be two TCD of different diameters. This is due to the direction of the rotation of the propeller. For most single screw Merchant ships, the propeller rotates in a clockwise direction when viewed from aft to forward part of the ship. It does make a difference to the TCD.

It should be observed in Figure 15.3 that at the beginning of the Port turning manoeuvre, the ship turns initially to Starboard. There are reasons for this. Forces acting on the rudder itself will cause this move at first to Starboard. Larger centrifugal forces acting on the ship's hull will then cause the vessel to move the ship on a course to Port as shown in this diagram.

Merchant ships usually turn in a circle having a diameter of about 3–4 times the length between perpendiculars (LBP). The larger the rudder, the smaller will be the TCD. During the TCD manoeuvre, the ship will experience transfer, advance, drift angles and angle of heel (see Figure 15.3).

The maximum angle of heel must be recorded. If the ship has Port rudder helm this final angle of heel will be to Starboard and vice versa. Again, this is due to centrifugal forces acting on the ship's hull.

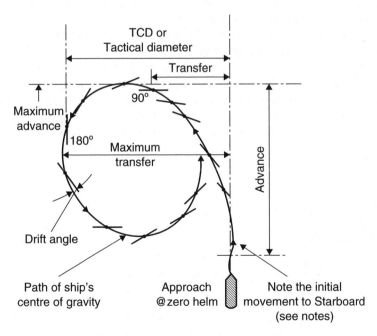

Fig. 15.3   TCD manoeuvres. Ship run at full speed with rudder helm 35°
P or S throughout this trial.

Ship model tests and Ship Trials have shown that the TCD does not change if this trial is run at speeds less than full speed. If these trials had been carried out in shallow waters, the TCD could have been *double* that measured in deep-water conditions.

Crash-stop manoeuvres

In these manoeuvres, the rudder helm is set at a fixed angle of zero. The procedure for Crash-stop trials is as follows:

1. Ship is at full speed. Order is then given on the bridge 'full ahead power to be shut off. Stop engines.' The ship will begin to slow down owing to frictional resistance on the shell plating and the underwater hull form (see Figure 15.4).
2. Propeller shaft speed drops until zero slip with zero thrust occurs.
3. After a period of time 't' dependent on type of machinery installed, reverse torque is then applied. The propeller slows down, stops and then begins to go on astern revolutions. As soon as negative slip is reached, astern characteristics are set in motion.
4. Astern torque must now be gradually increased until maximum astern torque value is reached. Figure 15.4 shows how this is now held until vessel's speed slows down to zero.

If Steam Turbine machinery has been fitted, then the 'full astern power' is about 40% of the 'full ahead power.'

If Diesel machinery had been installed, then the 'full astern power' is about 80% of the 'full ahead power.'

Consequently, ships fitted with Diesel machinery will have Crash stops that are comparatively less in distance and in time for the ship to come to a halt. Thus the type of main machinery installed is important for this manoeuvre. Ships fitted with Diesel machinery will have stopping distances of approximately 70% of those fitted with Steam Turbine machinery.

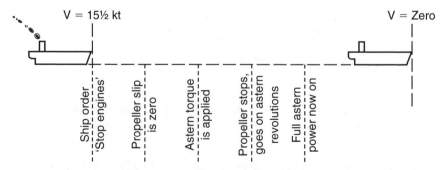

Fig. 15.4   Crash-stop manoeuvre programme of events.

For Figure 15.5 it is assumed that:
- No movement from 'rudder amidships' throughout these manoeuvres.
- They were all single screw Oil Tankers with a C_B of the order 0.800–0.825.
- Service speed at 'full ahead power' was 15.50–16.00 kt.
- All vessels were fitted with Steam Turbine machinery.

Lateral deviation (see Figure 15.6) ranged from 1/4 to 1/3 times the stopping distance (S). This can be to Port or be to Starboard. The lateral deviation is capricious. It is due to hydrodynamic imbalance.

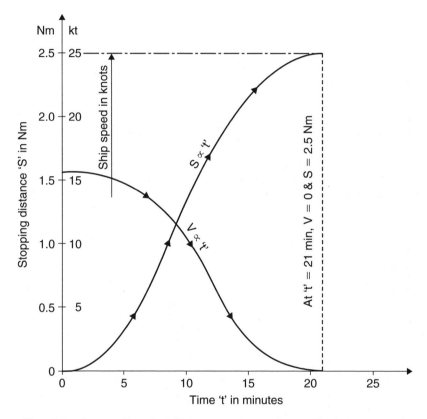

Fig. 15.5   Curves for a 215 000 tonnes dwt tanker on Crash-stop tests.

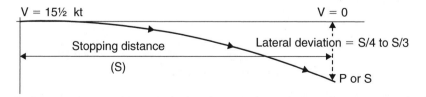

Fig. 15.6   Lateral deviation for Oil Tankers on Crash-stop tests.

Table 15.1 Stopping distances and times for a selection of Oil Tankers on Crash-stop manoeuvres in deep-water conditions

Oil Tanker dwt (tonnes)	Stopping distance (S) in nautical miles	S/LBP or lengths to stop each tanker	Time to stop each tanker (min)
50 000	1.49	11.5	11.0
100 000	1.60	12.5	12.0
150 000	1.89	13.5	15.0
200 000	2.36	14.5	19.5
250 000	3.00	15.5	25.0

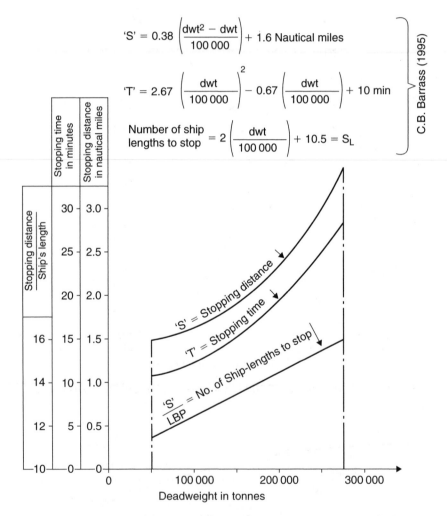

$$\text{'S'} = 0.38 \left(\frac{\text{dwt}^2 - \text{dwt}}{100\,000} \right) + 1.6 \text{ Nautical miles}$$

$$\text{'T'} = 2.67 \left(\frac{\text{dwt}}{100\,000} \right)^2 - 0.67 \left(\frac{\text{dwt}}{100\,000} \right) + 10 \text{ min}$$

$$\text{Number of ship lengths to stop} = 2 \left(\frac{\text{dwt}}{100\,000} \right) + 10.5 = S_L$$

C.B. Barrass (1995)

'S' = Stopping distance

'T' = Stopping time

$\frac{\text{'S'}}{\text{LBP}}$ = No. of Ship-lengths to stop

Fig. 15.7   Crash-stop information from measurements made on Oil Tankers fitted with Steam Turbine machinery.

Table 15.1 gives the Crash-stop results for a group of full size Shell Tanker Ltd Oil Tankers. These tankers ranged in deadweight (dwt) from about 50 000 to about 250 000 tonnes. The author plotted the reported results (see Figure 15.7) and then lifted mean values from these graphs (see Figure 15.7) to obtain the values shown in Table 15.1.

The author (in 1995) produced equations (see page 142) for the values in Table 15.1 and for the curves in Figure 15.7. It must be emphasised that these are for deep-water conditions.

If these tests had been carried out in shallow waters then the stopping distances and stopping times would be larger in value. Because of the additional entrained water effects and extra kinetic energy in shallow waters, the stopping distances can be up to 2.5 times those shown in Table 15.1.

Questions

1 An Oil Tanker has a dwt of 175 000 tonnes. She is undergoing Crash-stop trials in deep waters. Estimate her stopping distance in nautical miles and in the number of ship lengths. Proceed to estimate her stopping time in minutes.

2 Sketch the diagram for a ship's Zig-zag trial and label the important points on this diagram.

3 With the aid of a sketch, define the following four terms in a TCD manoeuvre:

    Advance, drift angle, tactical diameter and maximum transfer.

4 The stopping distance (S) for an Oil Tanker in deep waters on a Crash-stop test is 2.28 nautical miles. Estimate the transverse lateral deviation by the time this vessel has come to zero speed. If this test had been repeated in very shallow waters, what could have been the stopping distance?

Chapter 16

Ship Trials: residual trials

1. Anchor/cable/windlass trials.
2. Astern trials.
3. Rudder helm trials.
4. Transverse-thruster tests.
5. Hydraulic-fin stabiliser tests.
6. Bollard pull trials.
7. Navigation instrumentation checks.
8. Communications equipment testing.
9. Tank integrity tests.
10. Main and auxiliary power checks.
11. Lifeboat and release chutes release tests.
12. Accommodation checklists.

Anchor/cable/windlass trials (as per Lloyds Rules)

All anchors and chain cables are to be tested at establishments and on machines recognised by the committee and under the supervision of Lloyds Rules'(LR's) Surveyors or other Officers recognised by the committee. Test certificates showing particulars of weights of anchors or size and weight of cable and of the test loads applied are to be furnished and placed on board ship.

Anchor lowering and hoisting tests

Water depth to be at least 30 m. Ship forward speed to be at or very near to zero. Tests are made to examine the brake control as anchor is lowered to 30 m water depth. *Single* and *double* hoists are made, from 30 m water depth until clear of the water. Measurements are made of the water depth, length of chain below water level, chain speed (at least 9 m/min) of hoist, ammeter reading and hydraulic pressure. These hoists are to be made when ship speed is zero.

Windlass design and testing

A windlass of sufficient power and suitable for the size of the anchor chain is to be fitted to the ship. During trials on board ship, the windlass is shown to be capable of the following:

(a) for all specified design anchorage depths: raising the anchor from a depth of 82.5 to 27.5 m at a mean speed of 9 m/min;

(b) for specified design anchorage depths greater than 82.5 m: in addition to (a), raising the anchor from the specified design anchorage depth to a depth of 82.5 m at a mean speed of 3 m/min.

Where the depth of water in the trial area is inadequate, suitable equivalent simulating conditions will be considered as an alternative.

Windlass performance characteristics are based on the following assumptions:

(a) one cable lifter only is connected to the drive shaft,
(b) continuous duty and short-term pulls are measured at the cable lifter,
(c) brake tests are carried out with the brakes fully applied and the cable lifter declutched,
(d) the probability of declutching a cable lifter from the motor with its brake in the off position is minimised,
(e) hawse pipe efficiency is assumed to be 70%.

Where shipowners require equipment significantly in excess of LR requirements, it is their responsibility to specify increased windlass power.

The chain locker is to be of a capacity and depth adequate to provide an easy direct lead for the anchor cable into the chain pipes, when the anchor cable is fully stowed. Chain lockers fitted abaft the collision bulkhead are to be watertight and the space to be efficiently drained.

Astern trials

These trials are more important for smaller ships such as ferries and coasters that are constantly manoeuvring in and out of Ports. When 'full astern revolutions' are requested by the Officers on the bridge, the ship response is measured.

Characteristics such as rates of turn and times when going astern using rudder helm (can be up to 15° P or S) will also be noted.

Rudder helm trials

This trial measures the efficiency of the ship's main Steering Gear. The requirement is that the Steering Gear in the S.G. Compartment must be able to turn the ship's rudder from say, a helm of *35°Port* to a helm of *30° Starboard* in at least *28 sec*, when the ship is at *full service speed*.

Transverse-thruster propulsion unit trials

Units may be fitted forward in the Forward Peak tank or in the Aft Peak tank. Transverse movement of ship is measured when the forward speed of ship is 0, 2 and 4 kt. When the ship's forward speed is *zero*, it is expected that these T/T propulsion units are most effective.

'Hand-roll' test for hydraulic-fin stabilisers

Whilst at full speed, the vessel is hand rolled to about 10° P&S by activated feathering movements of the extended fins.

When the fins are changed over to a stabilising mode observations are then made and recorded of the damping effect, from the 10° angle of roll to the almost upright condition. The vessel will roll P&S until eventually angle of heel will decay to zero.

Feedback from ships in heavy sea conditions have shown that these hydraulic-fin stabilisers can dampen a roll of 30° P&S down to a roll of only 1.5° roll P&S when extended outboard and in stability mode. Very effective but expensive to have fitted and expensive to repair in the event of a breakdown.

Bollard pull trials

These trials are carried out mainly on Tugs and Voith–Schneider water tractors. The ship is connected to a shore-based bollard by a towing wire of at least twice the ship's length (see Figure 16.1). Other pre-requisites include:

- No tidal or current effects, with a maximum tidal speed to be 1 kt.
- Wind < force 3 or 4.
- Calm river or sea conditions with no swell.

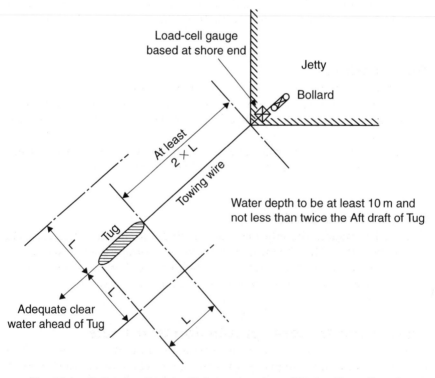

Fig. 16.1   Bollard pull trials. *Reference source*: 'Marine Propellers & Propulsion' by J.S. Carlton (Elsevier).

The actual bollard pull tests are as follows:

- Maximum bollard pull for 1 *min* at maximum input of ship's engine power.
- Steady bollard pull over a period of 5 *min*.
- Effective bollard pull, e.g. in open water conditions. In practice, this is usually approximated to 78% of the steady bollard pull after making due allowance for weather conditions.
- For Coastal Tugs, a typical bollard pull is in the range of 15–30 tonnes.
- For Ocean-going Tugs and Port Tugs, bollard pull is in the range of 30–115 tonnes. The high bollard pull of 115 tonnes was for a Tug built in 1996.

Navigation instrumentation checks

All radar and navigational systems are to be checked for accuracy. They may be satisfactory on shore, but need to be tested in sea going conditions.

Communications equipment testing

All radio and communications equipment including walkie talkies must be rechecked for 'at sea' conditions. If necessary they must be recalibrated.

Tank integrity tests

Cargo tanks and Ballast tanks must be checked to verify that there is no ingress of water through the sideshell. This is another check for seaworthiness. There must be no seepage through main tank bulkheads from one tank to another.

Main and auxiliary power checks

Full main, auxiliary and emergency power and hand steering checks must be carried out. There must be an arrangement for emergency lighting if the main power is lost. Another example is, if the steering gear machinery power fails, the ship must possess a hand control back-up system for obtaining rudder helm.

Lifeboat and release chutes release tests

Lifeboats, conventional and stern davit launched boats to be drop tested. Food and medical supplies to be verified as being to required standards.

Accommodation checklists

All internal sanitary systems to accommodation and public spaces to be tested. All accommodation ladders and gangways to be checked. Medical stocks to be at least up to International and IMO standards.

Chapter 17

Ship squat in open water and in confined channels

What exactly is ship squat?

When a ship proceeds through water, she pushes water ahead of her. In order not to leave a 'hole' in the water, this volume of water must return down the sides and under the bottom of the ship. The streamlines of return flow are speeded up under the ship. This causes a drop in pressure, resulting in the ship dropping vertically in the water.

As well as dropping vertically, the ship generally trims for'd or aft (see Figure 17.1). Ship squat thus is made up of two components, namely mean bodily sinkage plus a trimming effect. If the ship is on even keel when

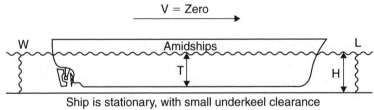

Ship is stationary, with small underkeel clearance
Trim ranges from being on even keel,
to being 1/500 by the stern

Ship Squat

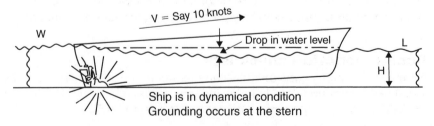

Ship is in dynamical condition
Grounding occurs at the stern

Fig. 17.1   Container vessel squatting at the stern.

static, the trimming effect depends on the ship type and C_B being considered. Also read later in this chapter, the detailed note on false drafts.

The overall decrease in the static underkeel clearance (ukc), for'd or aft, is called ship squat. It is *not* the difference between the draughts when stationary and the draughts when the ship is moving ahead.

If the ship moves forward at too great a speed when she is in shallow water, say where this static even-keel ukc is 1.0–1.5 m, then grounding due to excessive squat could occur at the bow or at the stern.

For full-form ships such as Supertankers or OBO vessels, grounding will occur *generally* at the bow. For fine-form vessels such as Passenger Liners or Container ships the grounding will generally occur at the stern. This is assuming that they are on *even keel* when stationary.

If C_B is >0.700, then maximum squat will occur at the bow.

If C_B is <0.700, then maximum squat will occur at the stern.

If C_B is very near to 0.700, then maximum squat will occur at the stern, amidships and at the bow. The squat will consist only of mean bodily sinkage, with no trimming effects.

It must be *generally*, because in the last two decades, several ship types have tended to be shorter in length between perpendiculars (LBP) and wider in Breadth Moulded (Br. Mld). This has lead to reported groundings due to ship squat at the bilge strakes at or near to amidships when rolling motions have been present.

Why has ship squat become so important in the last 40 years?

Ship squat has always existed on smaller and slower vessels when underway. These squats have only been a matter of centimetres and thus have been inconsequential.

However, from the mid-1960s to this new millennium, ship size steadily has grown until we have Supertankers of the order of 350 000 tonnes deadweight (dwt) and above. These Supertankers have almost out-grown the Ports they visit, resulting in small static even-keel ukc of only 1.0–1.5 m.

Alongside this development in ship size has been an increase in service speed on several ships, e.g. Container ships, where speeds have gradually increased from 16 up to about 25 kt.

Ship design has seen tremendous changes in the 1980s and 1990s. In Oil Tanker design we have the 'Jahre Viking' with a dwt of 564 739 tonnes and an LBP of 440 m. This is equivalent to the length of five football pitches.

In 2002, the biggest Container ship to date, namely the 'Hong Kong Express' came into service. She has a dwt of 82 800 tonnes, a service speed of 25.3 kt, an LBP of 304 m, Br. Mld of 42.8 m and a draft moulded of 13 m.

As the static ukc have decreased and as the service speeds have increased, ship squats have gradually increased. They can now be of the order of 1.50–1.75 m, which are of course by no means inconsequential.

Recent ship groundings

To help focus the mind on the dangers of excessive squat one only has to recall the grounding of these nine vessels in recent years.

Herald of Free Enterprise	RO-RO vessel at Zeebrugge	06/03/1987
QE11	Passenger Liner at New Orleans	07/08/1992
Sea Empress	Supertanker at Milford Haven	15/02/1996
Heidrun	Supertanker at Nantes	10/09/1996
Diamond Grace	260 000 tonnes dwt Very Large Crude Carrier (VLCC) at Tokyo Harbour	02/07/1997
Napoleon Bonaparte	Passenger Liner at Marseille	05/02/1999
Treguier	31 950 tonnes dwt Oil Tanker at Bordeaux	04/08/1999
Don Raul	37 000 tonnes Bulk Carrier at Pulluche, Chile	31/03/2001
Tasman Spirit	87 500 tonnes Oil Tanker at Karachi Harbour	27/07/2003

Department of Transport 'M' notices

In the UK, over the last 20 years the UK Department of Transport have shown their concern by issuing *four* 'M' notices concerning the problems of ship squat and accompanying problems in shallow water. These alert all Mariners to the associated dangers.

Signs that a ship has entered shallow water conditions can be one or more of the following:

1. Wave-making increases, especially at the forward end of the ship.
2. Ship becomes more sluggish to manoeuvre. A pilot's quote ... 'almost like being in porridge.'
3. Draught indicators on the bridge or echo sounders will indicate changes in the end draughts.
4. Propeller rpm indicator will show a decrease. If the ship is in 'open water' conditions, i.e. without breadth restrictions, this decrease may be up to 15% of the Service rpm in deep water. If the ship is in a confined channel, this decrease in rpm can be up to 20% of the service rpm.
5. There will be a drop in speed. If the ship is in open water conditions this decrease may be up to 30%. If the ship is in a confined channel such as a river or a canal then this decrease can be up to 60%.
6. The ship may start to vibrate suddenly. This is because of the entrained water effects causing the natural hull frequency to become resonant with another frequency associated with the vessel.

7. Any rolling, pitching and heaving motions will all be reduced as ship moves from deep water to shallow water conditions. This is because of the cushioning effects produced by the narrow layer of water under the bottom shell of the vessel.
8. The appearance of mud could suddenly show in the water around the ship's hull say in the event of passing over a raised shelf or a submerged wreck.
9. Turning circle diameter (TCD) increases. TCD in shallow water could increase 100%.
10. Stopping distances and stopping times increase, compared to when a vessel is in deep waters.
11. Rudder is less effective when a ship is in shallow waters.

What are the factors governing ship squat?

The main factor is ship speed V. Detailed analysis has shown that squat varies as speed to the power of 2.08. However, squat can be said to vary approximately with the speed squared. In other words, we can take as an example that if we halve the speed we quarter the squat. Put another way, if we double the speed we quadruple the squat!!

In this context, speed V is the ship's speed relative to the water. Effect of current/tide speed with or against the ship must therefore be taken into account.

Another important factor is the block coefficient C_B. Squat varies directly with C_B. Oil Tankers will therefore have comparatively more squat than Passenger Liners.

The Blockage Factor 'S' is another factor to consider (see Figure 17.2). This is the immersed cross-section of the ship's midship section divided by the cross-section of water within the canal or river. If the ship is in open water the width of influence of water can be calculated. This ranges from about 8.25 breadths for Supertankers, to about 9.50 breadths for General Cargo ships, to about 11.75 ship breadths for Container ships. See Chapter 18 for detailed notes on the 'width of influence.'

Water depth (H)/ship's draft (T) also affects ship squat. When H/T is 1.10–1.40, then squat varies as the reciprocal of H/T. Hence squat will vary as T/H.

The presence of another ship in a narrow river (passing, overtaking or simply moored) will also affect squat, so much so, that squats can *double in value* as they pass/cross the other vessel.

Squat formulae

Formulae have been developed that will be satisfactory for estimating maximum ships squats for vessels operating in confined channels and in open water conditions. These formulae are the results of analysing about 600 results. Some have been measured on ships and some on ship models. Some

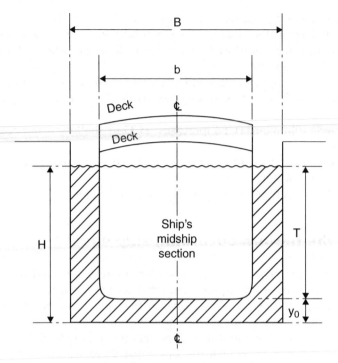

A_S = cross-section of ship at amidships = $b \times T$
A_C = cross-section of canal = $B \times H$ *or* 'B' $\times H$
$A_W = A_C - A_S$
y_0 = static underkeel clearance

Blockage factor = $\dfrac{A_S}{A_C}$ = S

'B' = $\dfrac{7.04}{C_B^{0.85}} \times b$

'Width of influence' = $\dfrac{\text{equivalent 'B'}}{\text{ship's breadth 'b'}}$ in open water conditions

V = speed of ship relative to water in knots
Blockage factor covers range of 0.10–0.266

Fig. 17.2   Ship in a canal in static condition.

of the empirical formulae developed are as follows:
    Let:

    b = breadth of ship
    H = depth of water
    C_B = block coefficient

CSA = cross-sectional area
B = breadth of river or canal
T = ship's even-keel static draft
V = ship speed relative to the water or current

$$\text{Let Blockage factor} = S = \frac{\text{CSA of ship}}{\text{CSA of river or canal}}$$

If ship is in open water conditions, then the formula for B becomes:

$$B = \frac{7.04}{C_B^{0.85}} \text{ ship breadths} \qquad \text{known as the 'width of influence'}$$

$$\text{Blockage factor} = S = \frac{b \times t}{B \times H}$$

$$\text{Maximum squat} = \delta_{max} = \frac{C_B \times S^{0.81} \times V^{2.08}}{20} \text{ m} \quad \text{for open water and confined channels}$$

Two short-cut formulae relative to the previous equation are:

$$\delta_{max} = \frac{C_B \times V^2}{100} \text{ m} \quad \text{for open water conditions only} \quad \text{with } \frac{H}{T} = 1.1\text{--}1.4$$

$$= \frac{C_B \times V^2}{50} \text{ m} \quad \text{for confined channels} \quad \text{where } S = 0.100\text{--}0.266$$

An 'S' value of 0.100 appertains to a very wide river, almost in open water conditions. An 'S' value of 0.226 appertains to a narrow river.
For a medium width of river,

$$\delta_{max} = K \times \frac{C_B \times V^2}{100} \text{ m} \qquad \text{for medium width rivers}$$

$$K = (6 \times \text{'S'}) + 0.40 \qquad \text{S is the blockage factor}$$

A worked example, showing how to predict maximum squat and how to determine the remaining ukc is shown later in this chapter. It illustrates the use of the more detailed formula and then compares the answer with the short-cut method.

These formulae have produced several graphs of maximum squat against ships speed V. One example of this is in Figure 17.3, for a 250 000 tonnes dwt Supertanker. Another example is in Figure 17.4, for a Container vessel having shallow water speeds up to 18 kt.

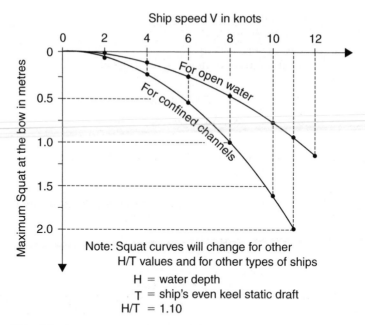

Fig. 17.3   Maximum squats against ship speed for a 250 000 tonnes dwt Supertanker.

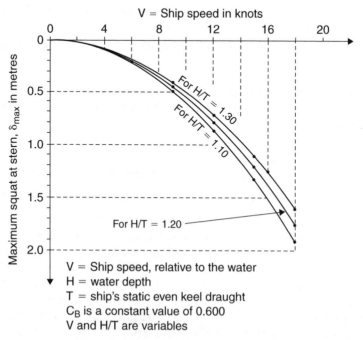

Fig. 17.4   Squats for Container vessels in open water when C_B is 0.600.

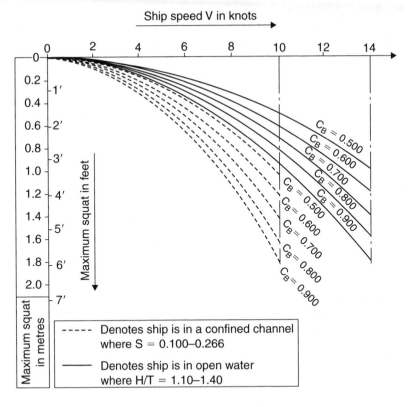

Fig. 17.5   Maximum ship squats in confined channels and in
open water conditions.

Ship type	Typical C_B, fully-loaded	Ship type	Typical C_B, fully-loaded
ULCC	0.850	General Cargo	0.700
Supertanker	0.825	Passenger Liner	0.625
Oil Tanker	0.800	Container ship	0.565
Bulk carrier	0.775–0.825	Coastal tug	0.500

Figure 17.5 shows the maximum squats for Merchant ships having C_B values from 0.500 up to 0.900, in open water and in confined channels. Three items of information are thus needed to use this diagram.

First, an idea of the ship's C_B value, secondly the selected speed V and thirdly to decide if the ship is in open water or in confined river/canal conditions. A quick graphical prediction of the maximum squat can then be made. The final decision to be made is whether the remaining ukc at the bow or the stern is safe enough. If it is not safe, then the selected speed prior to the ship's transit should be reduced.

Worked example 17.1

In shallow water conditions, for a fully loaded condition, a vessel has a C_B of 0.750. She was on even keel when static. Estimate the maximum squat at the bow when she has a speed of 10 kt in open water and when she is in a confined channel.

$$\delta_{max} = \frac{C_B \times V^2}{100} \text{ m} \qquad \text{for open water conditions only}$$

$$= \frac{0.750 \times 10^2}{100} = 0.75 \text{ m at the bow, because } C_B > 0.700$$

$$\delta_{max} = \frac{C_B \times V^2}{50} \text{ m} \qquad \text{for confined channels only}$$

$$= \frac{0.750 \times V^2}{50} = 1.50 \text{ m at the bow, because } C_B > 0.700$$

Worked example 17.2

If the ship in Worked example 17.1 was operating in a river giving a blockage factor of 0.175, then estimate the maximum squat as she proceeds at a forward speed of 10 kt.

For a medium with of river,

$$\delta_{max} = K \times \frac{C_B \times V^2}{100} \text{ m} \qquad \text{medium width rivers}$$

$$K = (6 \times \text{'S'}) + 0.40 \qquad \text{hence} \quad K = (6 \times 0.175) + 0.40 = 1.45$$

$$\text{Therefore} \quad \delta_{max} = 1.45 \times \frac{0.750 \times 10^2}{100}$$

$$= 1.09 \text{ m at the bow, because } C_B > 0.700$$

Worked example 17.3

A Supertanker is operating in open water conditions. Her Br. Mld is 55 m. Her C_B is 0.830, static even-keel draft (T) is 13.5 m and forward speed is 11 kt. The water depth (H) is 16 m. Calculate the maximum squat for this vessel by *two* methods and her minimum remaining ukc at this speed of 11 kt.

$$\text{Width of influence} = B = \frac{7.04}{C_B^{0.85}} \text{ ship breadths}$$

$$= \frac{7.04}{0.830^{0.85}} \times 55 = 8.248 \times 55$$

$$= 453.6 \text{ m}$$

This is an equivalent artificial width of river in open water conditions. Any greater width of water will give the same values for maximum squats for this vessel only.

$$\text{Blockage factor} = S = \frac{b \times t}{B \times H} = \frac{55 \times 13.5}{453.6 \times 16} = 0.102$$

Method 1 (more detailed method)

$$\text{Max squat} = \delta_{max} = \frac{C_B \times S^{0.81} \times V^{2.08}}{20} \text{ m} \qquad \text{for open water}$$

$$= \frac{0.830 \times 0.102^{0.81} \times 11^{2.08}}{20}$$

$$= 0.96 \text{ m at the bow, because } C_B > 0.700$$

Method 2 (short-cut method)

$$\delta_{max} = \frac{C_B \times V^2}{100} \text{ m} \qquad \text{for open water}$$

$$= 0.830 \times \frac{11^2}{100}$$

$$= 1.00 \text{ m at the bow, because } C_B > 0.700$$

This is slightly above the first answer, so is erring on the high and therefore safe side.

$$\text{Average maximum squat is } \frac{0.96 + 1.00}{2} = 0.98 \text{ m}$$

Hence remaining ukc at the bow $= H - T - \delta_{max}$

$$= 16.00 - 13.50 - 0.98$$

$$= 1.52 \text{ m @ } V = 11 \text{ kt}$$

Worked example 17.4

Use Figure 17.5 to estimate the maximum squats for Cargo–Passenger ship having a C_B of 0.650 and a forward speed of 8.00 kt, when she is in open water and when she is in a confined channel.

Procedure

Open water: At a speed of 8 kt, drop vertically down from the 'x' axis until midway between the *solid curves* for C_B values of 0.600 and 0.700. At this point move left to the 'y' axis and lift off the maximum squat value of 0.42 m.

Confined channel: At a speed of 8 kt, drop vertically down from the 'x' axis until midway between the *dotted curves* for C_B values of 0.600 and 0.700. At this point move left to the 'y' axis and lift off the maximum squat value of 0.84 m.

Both of these maximum squats will occur at the stern, because C_B is <0.700.

Ship squat for ships with static trim

So far, each ship has been assumed to be on *even keel* when static. For a given forward speed the maximum ship squat has been predicted. Based on the C_B, the ship will have this maximum squat at the bow, at the stern or right away along the length of the ship.

However, some ships will have *trim by the bow* or *trim by the stern* when they are stationary. This static trim will decide whereabouts the maximum squat will be located when the ship is underway.

Tests on ship models and from full size squat measurements have shown that:

1. If a ship has static trim by the stern when static, then when underway she will have a maximum squat (as previously estimated) at the stern. The ship will have dynamic trim in the same direction as the static trim. In other words, when underway she will have increased trim and could possibly go aground at the stern (see Figure 17.6).

   This is because streamlines under the vessel at the stern move faster than under the vessel at the bow. Cross-sectional area is less at the stern than under the bow. This causes a greater suction at the stern. Vessel trims by the stern. In hydraulics, it is known as the venturi effect.

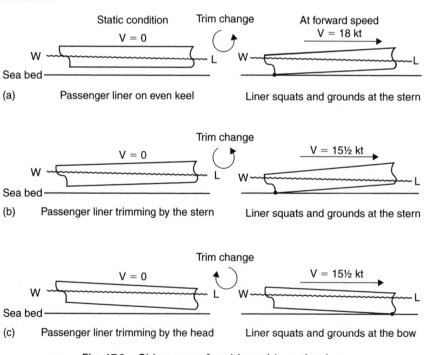

Fig. 17.6   Ship squats for ships with static trim.

2. If a ship has static trim by the bow when static, then when underway she will have a maximum squat (as previously estimated) at the bow. The ship will have dynamic trim in the same direction as the static trim (see Figure 17.6). In other words, when underway she will have increased trim and could possibly go aground at the bow. The *Herald of Free Enterprise* grounding was a prime example of this trimming by the bow when at forward speed.

Note of caution: Some Masters on Oil Tankers trim their vessels by the stern before going into shallow waters. They believe that full-form vessels trim by the bow when underway. In doing so they believe that their ship will level out at even keel when at forward speed. This does not happen!! What does happen is as per Figure 17.6(b).

Squats at both ends of a vessel in open water

Earlier notes have indicated how to predict the maximum squat for ships in shallow waters. They are for open water conditions and where the ship when static on even keel. Three questions that often arise:

- What is the squat at the other end of the vessel (o/e)?
- What is the mean bodily sinkage (mbs)?
- What is the dynamic trim (t) at the forward speed of the ship?

Research has shown that the answer to these questions is all three are linked with the maximum ship squat δ_{max}.

where $\delta_{max} = C_B \times \dfrac{V^2}{100}$ m in open water conditions

$$\text{Squat at the other end} = K_{o/e} \times \delta_{max} \text{ m}$$

$$\text{Mean bodily sinkage} = K_{mbs} \times \delta_{max} \text{ m}$$

$$\text{Dynamic trim} = K_t \times \delta_{max} \text{ m}$$

$$K_{o/e} = 1 - 40(0.700 - C_B)^2$$

$$K_{mbs} = 1 - 20(0.700 - C_B)^2$$

$$K_t = 40(0.700 - C_B)^2$$

Figure 17.7 shows graphs for these K values for C_B ranging from 0.550 up to 0.850.

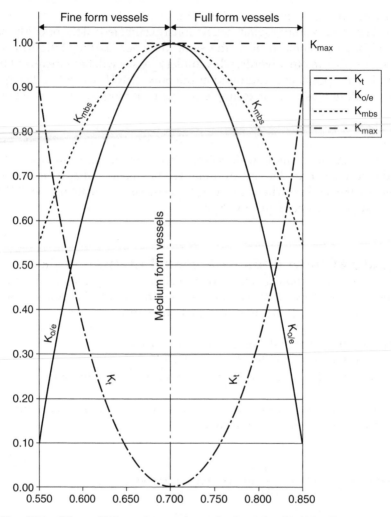

Fig. 17.7   'K' co-efficients for squats at both ends of a ship. Open water conditions with static even-keel drafts.

Worked example 17.5

A Bulk Carrier has a C_B of 0.815 and a static even-keel draft of 10.75 m in open water conditions. She is in a depth of water of 11.83 m. Calculate the squat at the bow and the stern, the mean bodily sinkage and the dynamic trim when the forward speed is 9.50 kt.

$$\delta_{max} = C_B \times \frac{V^2}{100} \text{ m in open water condtions}$$

$$= 0.815 \times 9.50 \times \frac{9.50}{100}$$

$$= 0.74 \text{ m at the } bow, \text{ because } C_B > 0.700$$

Squat at the other end, the *stern* $= K_{o/e} \times \delta_{max} = \{1 - 40(0.700 - C_B)^2\} \times \delta_{max}$

$$= \{1 - 40(0.700 - 0.815)^2\} \times 0.74 = 0.35\,m$$

Mean bodily sinkage $= K_{mbs} \times \delta_{max}\,m = \{1 - 20(0.700 - C_B)^2\} \times \delta_{max}$

$$= \{1 - 20(0.700 - 0.815)^2\} \times 0.74 = 0.545\,m$$

Dynamic trim $= 40(0.700 - C_B)^2 \times \delta_{max} = 40(0.700 \times 0.815)^2 \times 0.74$

$$= 0.39\,m \quad \text{by the bow, because the greater squat}$$
$$\text{occurred at the bow.}$$

Checks:

Dynamic trim $=$ maximum squat (at bow) $-$ squat at other end (at stern)

$$= 0.74 - 0.35 = 0.39\,m \text{ by the bow, this agrees spot on with}$$
$$\text{previous answer}$$

Mean bodily sinkage $= \dfrac{\overset{\text{Maximum squat}}{\underset{\text{(at bow)}}{}} - \overset{\text{Squat at other end}}{\underset{\text{(at stern)}}{}}}{2}$

$$= \dfrac{0.74 + 0.35}{2} = 0.545\ m \quad \text{this agrees spot on}$$
$$\text{with previous answer}$$

Thus, this Bulk Carrier will have a maximum squat at the bow of 0.74 m with a squat at the stern of 0.35 m. The mean bodily sinkage will be 0.545 m and the dynamic trim will be 0.39 m by the bow.

Remaining ukc under the bow $= H - T - \delta_{max} = 11.83 - 10.75 - 0.74$
$$= 0.34\,m$$

Remaining ukc under the stern $= H - T - \delta_{stern} = 11.83 - 10.75 - 0.35$
$$= 0.73\,m$$

Procedures for reducing ship squat

1. Reduce the mean draft of the vessel if possible by the discharge of water ballast. This causes two reductions in one:
   (a) At the lower draft, the block coefficient C_B will be slightly lower in value, although with Passenger Liners it will not make for a significant reduction. This is because of the boot-topping being a lot less than for many other ship types.
   (b) At the lower draft, for a similar water depth, the H/T will be higher in value. It has been shown that higher H/T values lead to smaller squat values.
2. Move the vessel into deeper water depths. For a similar mean ship draft, H/T will increase, leading again to a decrease in ship squat.
3. When in a river if possible, avoid interaction effects from nearby moving ships or with adjacent riverbanks. A greater width of water will lead to less ship squat unless the vessel is outside her width of influence.

4. The quickest and most effective way to reduce squat is to reduce the speed of the ship. Remember, halving the speed will quarter the squat.

False drafts

If a moored ship's drafts are read at a quayside when there is an ebb tide of say 4 kt then the draft readings will be false. They will be incorrect because the ebb tide will have caused a mean bodily sinkage and trimming effects. In many respects this is similar to the ship moving forward at a speed of 4 kt. It is actually a case of *the squatting of a static ship*.

It will appear that the ship has more tonnes displacement than she actually has. If a Marine Draft Survey is carried out at the next Port of Call (with zero tide speed), there will be a deficiency in the displacement 'constant.' Obviously, larger ships such as Supertankers and Passenger Liners will have greater errors in displacement predictions.

The deficiency in displacement will be the mean bodily sinkage times the tonnes per centimetre immersion (TPC). For a large Passenger Liner, loading up at a quayside, at an H/T of 1.25 in an ebb tide of 4 kt, this deficiency in displacement could be of the order of 1250 tonnes.

Summary

In conclusion, it can be stated that if we can predict the maximum ship squat for a given situation then the following advantages can be gained:

1. The ship operator will know which speed to reduce to in order to ensure the safety of his/her vessel. This could save the cost of a very large repair bill. It has been reported in technical press that the repair bill for the QEII was *$13 million* ... plus an estimate for lost Passenger bookings of *$50 million*!!

   In Lloyds lists, the repair bill for the 'Sea Empress' had been estimated to be in the region of *$28 million*. ... In May 1997, the repairs to the 'Sea Empress' were completed at Harland & Wolff Ltd of Belfast, for a reported cost of *£20 million*. Rate of exchange in May 1997 was of the order of £1 = $1.55. She was then renamed the 'Sea Spirit.'

2. The ship officers could load the ship up an extra few centimetres (except of course where load-line limits would be exceeded). If a 100 000 tonnes dwt Tanker is loaded by an extra 30 cm or an SD14 General Cargo ship is loaded by an extra 20 cm, the effect is an extra 3% onto their dwt. This gives these ships extra earning capacity.

3. If the ship grounds due to excessive squatting in shallow water, then apart from the large repair bill, there is the time the ship is 'out of service'. Being 'out of service' is indeed very costly because loss of earnings can be as high as £100 000 per day.

4. When a vessel goes aground there is always a possibility of leakage of oil resulting in compensation claims for oil pollution and fees for clean-up operations following the incident. These costs eventually may have to be paid for by the shipowner.

These last four paragraphs illustrate very clearly that not knowing about ship squat can prove to be very costly indeed. Remember, in a Marine Court hearing, ignorance is not acceptable as a legitimate excuse!!

Summarising, it can be stated that because maximum ship squat can now be predicted, it has removed the 'grey area' surrounding the phenomenon. In the past, ship pilots have used 'trial and error', 'rule of thumb' and years of experience to bring their vessels safely in and out of Port.

Empirical formulae quoted in this study, plus squat curves modified and refined by the author over a period of over 30 years research give *firmer guidelines*. By maintaining the ship's trading availability a shipowner's profits margins are not decreased. More important still, this report can help prevent loss of life as occurred with the 'Herald or Free Enterprise' grounding.

It should be remembered that the quickest method for reducing the danger of grounding due to ship squat is to … *reduce the ship's speed*.

'Prevention is better than cure' … less worry and much cheaper!!

Questions

1 Define 'ship squat' and 'blockage factor.'
2 List five signs that a ship has entered shallow waters.
3 For a ship the C_B is 0.765, blockage factor is 0.248 and speed is 9.75 kt. If this vessel had been on even keel when static then calculate the maximum squat in shallow water and suggest with reasoning whereabouts it will occur.
4 A vessel has a block coefficient of 0.785. When static she was on even keel. Calculate the maximum squat when she proceeds at a forward speed of 9.50 kt in open water where the H/T is 1.10.
5 A Passenger Liner moves at a forward speed of 12 kt an shallow water. Estimate her maximum squat if her C_B is 0.618 when she is in:
   (a) open water conditions and (b) confined channel conditions.
6 What are the advantages to ship operators of knowing how to predict ship squat in open waters and in confined channels?
7 Give two reasons why ship squat is more important today than say 40 years ago.
8 Discuss how the value of the C_B affects the trim component of squat for vessels that are on even keel when stationary.
9 A Container vessel has a static even-keel draft of 12.00 m and a C_B of 0.585 in open water conditions. She is in shallow waters. Calculate the squat at the bow and the stern, the mean bodily sinkage and the dynamic trim when the forward speed is 12.00 kt.

Chapter 18

Reduced ship speed and decreased propeller revolutions in shallow waters

A ship that is in shallow water but has no breadth restrictions is said to be in restricted waters or in open water conditions.

A ship that is in shallow water and has breadth restrictions is said to be in confined channel conditions.

Width of influence

If a ship is in open water conditions, there is an artificial boundary Port and Starboard, parallel to her centreline, beyond which there are no changes in ship speed, ship resistance or in ship squat (see Figure 18.1). This artificial boundary is known as a 'width of influence' denoted by (F_B).

The value of F_B depends on the type of ship and the block coefficient. Inside this width of influence when moving ahead, the ship will experience

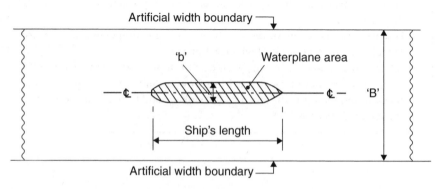

Fig. 18.1   Width of influence for Merchant ships.

Table 18.1 Widths and depths of influence for several Merchant ships

Ship type	Approximate C_B	Width of influence, F_B	Depth of influence, F_D
ULCCs	0.850	8.08 × B	5.48 × T
VLCCs	0.825	8.29 × B	5.70 × T
Oil Tankers	0.800	8.51 × B	5.93 × T
Bulk Carriers (small)	0.775	8.74 × B	6.18 × T
General Cargo ships	0.700	9.53 × B	7.06 × T
Passenger Liners	0.625	10.50 × B	8.18 × T
Container ships	0.575	11.27 × B	9.12 × T
Tugs	0.500	12.69 × B	10.93 × T

ULCCs = Ultra Large Crude carriers, VLCCs = Very Large Crude Carriers, B = Br. Mld and T = ship's static even keel draft.

a loss of speed and a decrease in propeller revolutions. The ship will also experience increased squat.

Let H = depth of water in metres.
Let T = ship's static even keel draft in metres.

After experiments using an electrical analogue and mathematical investigations into measured ship squats, the author is able to state that:

$$\text{Width of influence} = F_B = \frac{7.04}{C_B^{0.85}} \times \text{Breadth Moulded (Br. Mld) m}$$

Note: This formula only applies for H/T range of 1.10–1.40.

Table 18.1 shows the width of influence for several Merchant ships. The ship will perform differently when inside this width of influence, hence the name.

Depth of influence

There is also a 'depth of influence' (F_D) (see Figure 18.2). F_D will depend upon the type of ship and the block coefficient. Over the years, several maritime researchers such as Yamaguchi, Baker, Todd, Lackenby, Rawson and Tupper have suggested values for F_D.

The author has analysed their results together with his own and suggest that:

$$\text{Depth of influence} = F_D = \frac{4.44}{C_B^{1.3}} \times \text{Draft Mld m}$$

Table 18.1 shows the depth of influence for several Merchant ships. The ship will perform differently when below this depth of influence, hence the name.

The value of F_D depends on the type of ship and the block coefficient. Equal to and above this value the ship will be in deep water conditions. For a particular input of engine power, the ship speed, propeller revolutions and ship squat will not change. All of these values will be asymptotic.

Below this depth of influence the ship will be in shallow water. When moving ahead, the ship will experience a loss of speed and a decrease in propeller revolutions. The ship will also experience increased squat. In shallow water, the vessel will become more sluggish to manoeuvre.

Figures 18.3 and 18.4 show graphs of the width of influence and the depth of influence coefficients, against the block coefficient.

Figures 18.5 and 18.6 show the width of influence in terms of metres for several types of Merchant ships.

Figures 18.7 and 18.8 show the depth of influence in terms of metres for several types of Merchant ships.

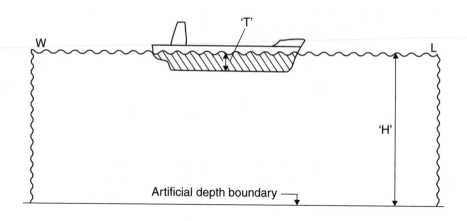

$$\text{Blockage factor} = \frac{\text{Ship's cross-sectional area at amidships}}{\text{Cross-section of 'channel'}}$$

$$\text{Therefore} \quad S = \frac{b \times T}{B \times H}$$

$$C_B = \frac{\text{Volume of displacement}}{L \times b \times T}$$

$$C_W = \frac{\text{Waterplane area}}{L \times b}$$

Fig. 18.2   Depth of influence for Merchant ships.

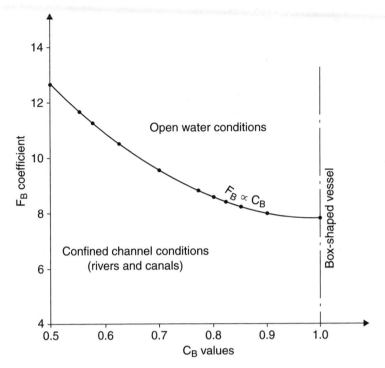

Fig. 18.3   Widths of influence coefficients.

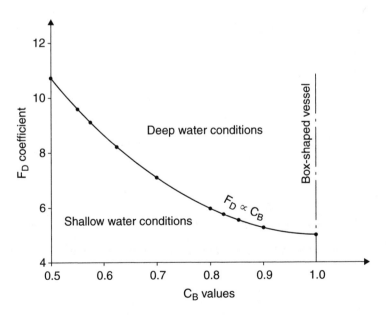

Fig. 18.4   Depths of influence coefficients.

b   = Breadth Mld of ship in metres
F_B = Width of influence in metres
C_B = Ship's block coefficient
CCC = Confined channel conditions
OWC = Open water conditions

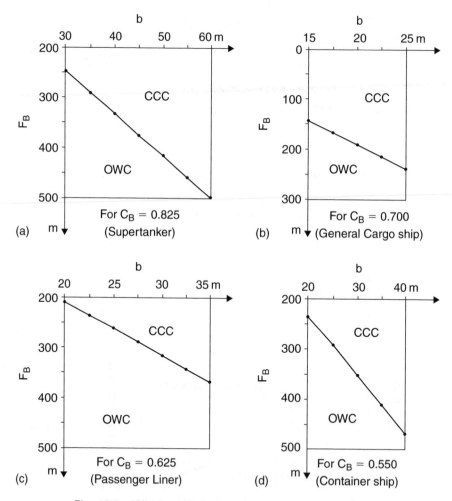

Fig. 18.5   Widths of influence for various types of ship.

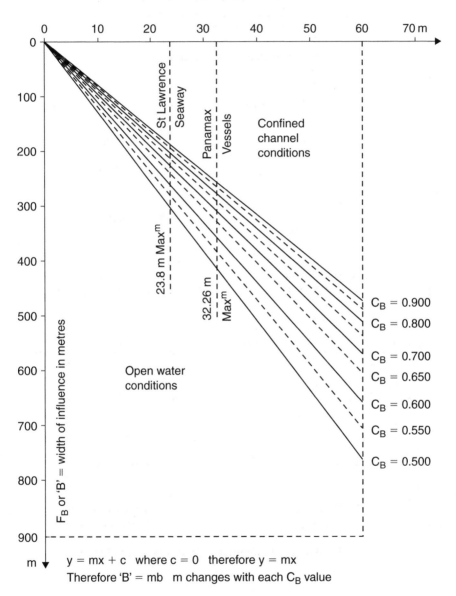

Fig. 18.6   Widths of influence for various values of C_B.

T = Static even keel draught in metres
F_D = Depth of influence in metres
C_B = Ship's block coefficient
SWC = Shallow water conditions
DWC = Deep water conditions

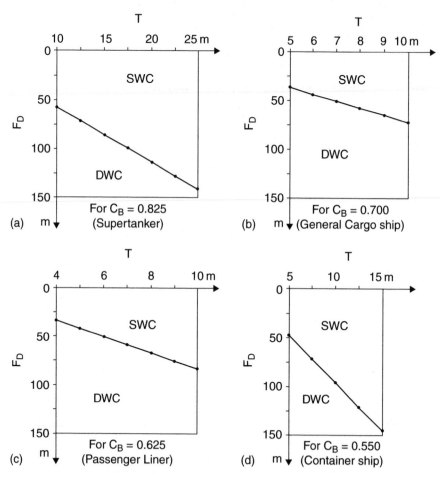

Fig. 18.7   Depths of influence for various types of ship.

C_B = Ship's block coefficient

'T' = Ship's static even keel draught in metres

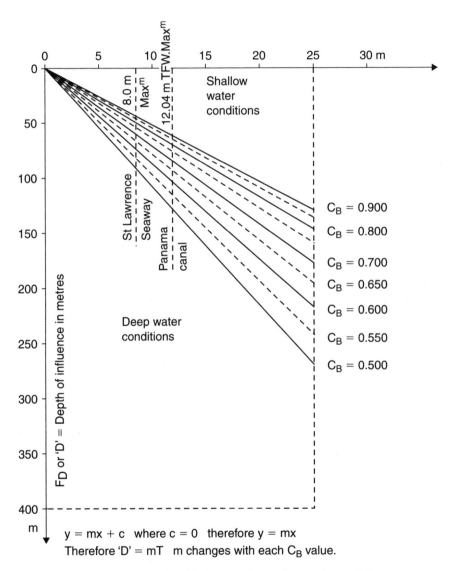

$y = mx + c$ where $c = 0$ therefore $y = mx$

Therefore 'D' = mT m changes with each C_B value.

Fig. 18.8   Depths of influence for various values of C_B.

Worked example 18.1

Calculate the width of influence and the depth of influence (in metres) for a vessel of 20 m Br. Mld, 8 m Draft Mld and a block coefficient of 0.725.

$$\text{Width of influence} = F_B = \frac{7.04}{C_B^{0.85}} \times \text{Br. Mld m}$$

$$= \frac{7.04}{0.725^{0.85}} \times 20$$

$$= 9.253 \times 20 = 185 \text{ m}$$

Equal to and greater than 185 m width of water, this ship is in open water conditions.

$$\text{Depth of influence} = F_D = \frac{4.44}{C_B^{1.3}} \times \text{Draft Mld m}$$

$$= \frac{4.44}{0.725^{1.3}} \times 8$$

$$= 6.744 \times 8 = 54 \text{ m}$$

Equal to and greater than 54 m depth of water, this ship is in deep water conditions.

Worked example 18.2

An Oil Tanker is on even keel in open water. Br.Mld is 50 m. C_B is 0.810. H/T is 1.20. Calculate her width of influence in metres and her maximum squat when her forward speed is 10 kt.

$$\text{Width of influence} = F_B = \frac{7.04}{C_B^{0.85}} \times \text{Br. Mld m}$$

$$= \frac{7.04}{0.810^{0.85}} \times 50 = 8.421 \times 50 = 421 \text{ m}$$

$$\text{Blockage factor} = S = \frac{b \times T}{B \times H} = \frac{1}{8.421 \times 1.20}$$

$$\text{Hence } S = 0.099$$

$$\text{Maximum squat} = \delta_{max} = \frac{C_B \times S^{0.81} \times V^{2.08}}{20} \text{ m}$$

$$= \frac{0.810 \times 0.099^{0.81} \times 10^{2.08}}{20}$$

$$= \frac{0.810 \times 0.1536 \times 120.23}{20}$$

$$= 0.75 \text{ m at the bow, because } C_B > 0.700$$

Note that if width of water had been 422, 500, 1000, 5000 m or greater, then the maximum squat would still have been 0.75 m. It would not have changed. If the width of water had been less than the 421 m width of influence, then the squat would have been more than this calculated value of 0.75 m.

Loss of speed and decrease in propeller revolutions for ships in shallow water

As previously stated, when a ship operates in shallow water her speed and propeller revolutions decrease. For the same input of engine power, her performance is not as good as when she is in deep water.

The reasons for these decreases are:

- The ship produces more waves. This produces more wave-making resistance thus causing extra drag to the vessel.
- Dynamical forces, emanating from the bottom shell travel downwards to the river or the seabed and reflect back onto the underside of the vessel.
- Due to increased turbulence at the aft end if the ship, the propeller efficiency, the propeller rpm and the Delivered power are all reduced.

The amounts that these the speed and propeller revolutions reduce will depend mainly on the following characteristics:

- Type of ship.
- Proportion of water depth (H)/(static mean draft of ship (T) i.e. the H/T value).
- Blockage factor (S) when the ship is stationary.

It will be shown later that these decreases in ship performance are exacerbated at low H/T values and at high blockage factors.

The Waginingen Model Test Tank conducted a series of tests into analysing these reductions of speed and rpm. These ship model tests represented Great Lakers, General Cargo ships, Container vessels, Passenger Liners in a seaway.

A.D. Watt reproduced a diagram of their findings showing how the percentage of ship speed, the percentage of propeller revolutions, H/T and blockage factors were inter-related.

The author considered these results only to when these ship models were simulated to be operating at 91% of maximum obtainable power. This was to more accurately simulate each ship's maximum continuous power (MCR rating).

H/T considerations

Figure 18.9 shows the percentage of ship speed relative to the full service speed in deep water, against the percentage of propeller rpm relative to the full rpm in deep water, in conjunction with a set of H/T values in open water conditions.

Each type of ship, in *deep water*, has been given a value of 'x' for H/T. The value of 'x' can be calculated using the previously given formulae for widths and depths of influence.

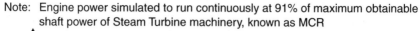

Note: Engine power simulated to run continuously at 91% of maximum obtainable shaft power of Steam Turbine machinery, known as MCR

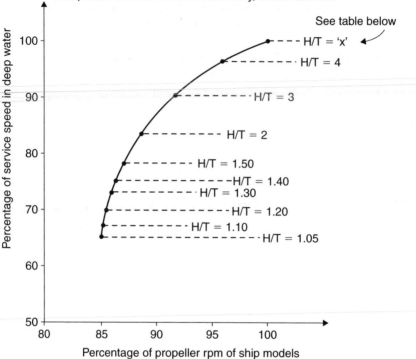

'x' is the H/T value, below which, each ship is in shallow water conditions.

At H/T values ⩾ x, each ship is in deep water conditions, with no percentage losses in speed or propeller rpm, for similar input of engine power

Fig. 18.9   Loss of speed and rpm in open water conditions.

Ship type	x = H/T	Ship type	x = H/T
Box-shaped vessels	4.96	General Cargo ships	7.07
ULCCs	5.20	Passenger ships	8.25
Supertankers	5.68	Container ships	9.70
Oil Tankers	5.91	Tugs	10.94

Blockage factor considerations

Figure 18.10 shows a graph with the axis similar to Figure 18.9 but both variables are related to the blockage factor 'S.' As can be seen, 'S' ranges from 0.200 to 0.275. Being so, they are therefore representative of confined channel conditions in a river or in a canal.

Figure 18.11 shows a cross-plot of information given in Figure 18.9. Clearly it can be observed that the percentage loss of service speed increases as H/T approaches the limiting value of unity. This changes parabolically. The next step was to derive an equation for this curve.

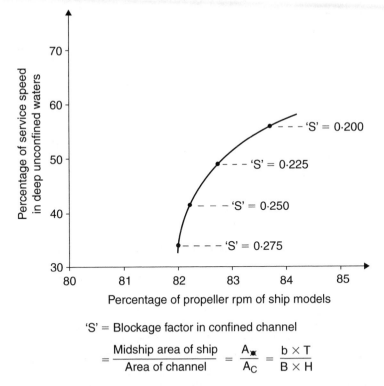

'S' = Blockage factor in confined channel

$$= \frac{\text{Midship area of ship}}{\text{Area of channel}} = \frac{A_{\bullet}}{A_C} = \frac{b \times T}{B \times H}$$

Fig. 18.10   Loss of ship-model speed and propeller rpm in confined channels.

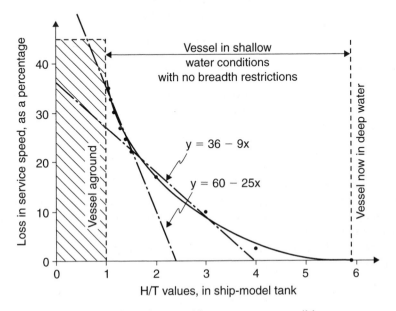

Fig. 18.11   Loss in speed in open water conditions.

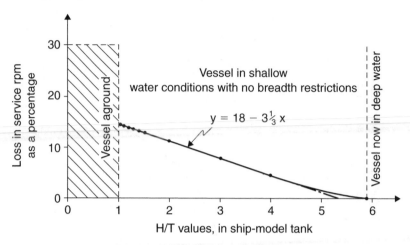

Fig. 18.12　Loss in propeller rpm in open water conditions.

However, it is possible to simplify this equation into two equations to cover first a range of H/T from 1.10 to 1.50 and secondly a range of H/T values from 1.50 to 3.00. The first range was selected by the author because it covered more dangerous situations leading to possible groundings. The second range was one that could lead to groundings at high speed, but the probability was comparatively decreased.

The equations produced by the author of this book were:

% loss in speed = 60 − (25 × H/T)　　　for an H/T of 1.10–1.50

% loss in speed = 36 − (9 × H/T)　　　for an H/T of 1.50–3.00

Figure 18.12 shows a cross-plot of information given in Figure 18.9 of losses in service propeller rpm against H/T. This time the graph is linear, the equation for which is:

% loss in propeller rpm = 18 − (10/3 × H/T)　for an H/T of 1.10–3.00

Figure 18.13 shows a cross-plot of information given in Figure 18.10 of losses of service speed against blockage factors. Figure 18.14 illustrates a cross-plot of information given in Figure 18.10 of losses of service rpm against blockage factors.

Figures 18.13 and 18.14 produced linear graphs, the equations for which are:

% loss of speed = (300 × S) − 16.5　　　　for S = 0.200–0.275.

% loss of propeller rpm = (24 × S) + 11.6　for S = 0.200–0.275.

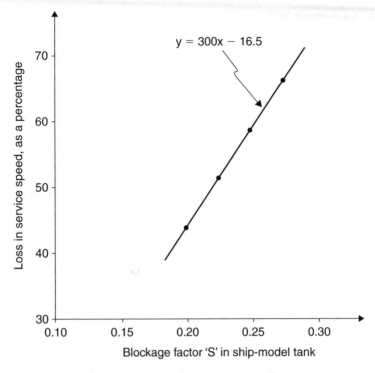

Fig. 18.13   Loss in ship-model speed in confined channels.

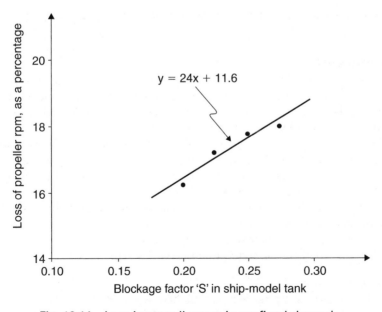

Fig. 18.14   Loss in propeller rpm in confined channels.

Worked example 18.3

In deep water, a ship has a shaft power of 5000 kW a service speed of 15 kt with propeller revolutions of 110 rpm. Calculate the ship speed and propeller revolutions in open water if the H/T is reduced to 1.15.

$$\% \text{ loss in speed} = 60 - (25 \times H/T) \quad \text{for an H/T of } 1.10\text{--}1.50$$

$$= 60 - (25 \times 1.15) = 31.25\%$$

Thus, ship speed $= (100\% - 31.25\%) \times 15 = 10.31$ kt   @ H/T of 1.15

$$\% \text{ loss in propeller rpm} = 18 - (10/3 \times H/T) \qquad \text{for an H/T of } 1.10\text{--}3.00$$

$$= 18 - (10/3 \times 1.15) = 14.17\%$$

Thus, propeller rpm $= (100\% - 14.17\%) \times 110 = 94$ rpm   @ H/T of 1.15

Worked example 18.4

In deep water, a ship has a shaft power of 5000 kW a service speed of 15 kt with propeller revolutions of 110 rpm. Calculate the ship speed and propeller revolutions in a river where the blockage factor S is 0.240.

$$\% \text{ loss of speed} = (300 \times S) - 16.5 \qquad \text{for S} = 0.200\text{--}0.275$$

$$= (300 \times 0.240) - 16.5 = 55.5\%$$

Thus ship speed $= (100\% - 55.5\%) \times 15 = 6.68$ kt   @ S = 0.240

$$\% \text{ loss of propeller rpm} = (24 \times S) + 11.6 \qquad \text{for S} = 0.200\text{--}0.275$$

$$= (24 \times 0.240) + 11.6 = 17.36\%$$

Thus, propeller rpm $= (100\% - 17.36\%) \times 110 = 91$ rpm   @ S = 0.240

Conclusions

It should be remembered that the Wageningen studies are for ship models only. It can easily be envisaged that ship owners are most reluctant to run their vessels in very low conditions. When H/T is 1.05–1.10, full size measurements and comparisons are few in number!!

There may be slight differences in speed losses and propeller rpm losses when scaled up from the ship models to full size vessels. Consequently, these graphs and suggested equations by the author should be treated as indications of the losses involved.

Some full size losses of speed and propeller revolutions however have been measured. They verify these equations to be satisfactory for shipboard personnel to use with their ships in shallow waters.

Summing up, it can be stated that in shallow waters:

For open water conditions
- Ship speed can decrease by about 30%, when H/T is 1.10–1.40.
- Propeller rpm can decrease by about 15%, when H/T is 1.10–1.40.

For confined channels
- Ship speed can decrease by 44–66%, when S = 0.200–0.275.
- Propeller rpm can decrease by 16–18%, when S = 0.200–0.275.

When the 'Sea Empress' went aground at Milford Haven in February 1996, first reports were that there had been a malfunction of the ship's machinery. The loss of speed, decrease in propeller revolutions and subsequent grounding were all blamed on the machinery.

However, as this chapter as shown, it was not due to a mechanical failure in the engine room. It was the hydrodynamic effects on the ship's hull that caused changes in the ship performance. Subsequent inspections by marine engineers later proved the machinery of the 'Sea Empress' to be in fine working order before and after the incident.

Questions

1 Define the terms, width of influence (F_B) and the depth of influence (F_D). Show how they are both linked with the ship's block coefficient.

2 Calculate the width of influence and the depth of influence for a RO-RO vessel having a Br.Mld of 31.5 m and a Draft Mld of 6.05 m with C_B of 0.582.

3 A vessel has a static even keel draft of 8.25 m in sea water. The water depth is 10 m. In deep water the service speed would be 15 kt. Estimate the loss of speed as a percentage and in knots, when operating in these shallow waters.

4 A propeller revolutions in deep water are 110 rpm. Calculate the loss in propeller revolutions when she enters a canal where the blockage factor is 0.225.

Chapter 19

The phenomena of Interaction of ships in confined waters

What exactly is Interaction?

Interaction occurs when a ship comes too close to another ship or too close to a river or canal bank. As ships have increased in size (especially in breadth moulded (Br. Mld)), Interaction has become very important to consider. In February 1998, the Marine Safety Agency (MSA) issued a Marine Guidance note 'Dangers of Interaction', alerting owners, masters, pilots and Tug-masters on this topic.

Interaction can result in one or more of the following characteristics:

1. If two ships are on a passing or overtaking situation in a river the squats of both vessels could be doubled when both amidships are directly in line.
2. When they are directly in line each ship will develop an angle of heel and the smaller ship will be drawn bodily towards the larger vessel.
3. Both ships could lose steerage efficiency and alter course without change in rudder helm.
4. The smaller ship may suddenly veer off course and head into the adjacent river bank.
5. The smaller ship could veer into the side of the larger ship or worse still be drawn across the bows of the larger vessel, bowled over and capsized.

In other words there is:

(a) a ship to ground Interaction,
(b) a ship to ship Interaction,
(c) a ship to shore Interaction.

What causes these effects of Interaction? The answer lies in the pressure bulbs that exist around the hull form of a moving ship model or a moving ship. See Figure 19.1, which shows these pressure bulbs in plan view. Figure 19.2 shows these pressure bulbs in profile view.

Note: Ship is moving ahead at velocity 'V'

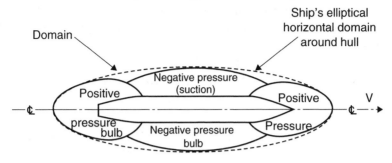

Fig. 19.1   Pressure distribution around ship's hull (not drawn to scale).

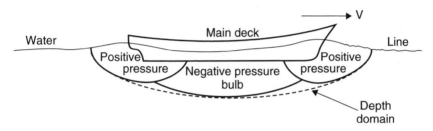

Fig. 19.2   Pressure bulbs beneath a moving ship (not drawn to scale).

As soon as a vessel moves from rest, hydrodynamics produce the shown positive and negative pressure bulbs. For ships with greater parallel body such as Oil Tankers, these negative bulbs will be comparatively longer in length. When ship is stationary in water of zero current speed these bulbs disappear.

Note the elliptical domain in Figure 19.1 that encloses the vessel and these pressure bulbs. This domain is very important. When the domain of one vessel interfaces with the domain of another vessel then Interaction effects will occur. Effects of Interaction are increased when ships are operating in shallow waters.

Ship to ground (squat) Interaction

In a report on measured ship squats in the St Lawrence seaway, A.D. Watt stated 'meeting and passing in a channel also has an effect on squat. It was found that when two ships were moving at the low speed of 5 kt that squat increased up to double the normal value. At higher speeds the squat when passing was in the region of one and a half times the normal value'. Unfortunately, no data relating to ship types, gaps between ships, blockage factors, etc. accompanied this statement.

Thus, at speeds of the order of 5 kt the squat increase is +100% whilst at higher speeds say 10 kt this increase is +50%. Figure 19.3 illustrates this passing manoeuvre. Figure 19.4 interprets the percentages given in previous paragraph.

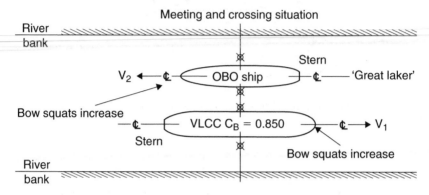

Fig. 19.3   Amidships (⊠) of Very Large Crude Carrier (VLCC) directly in line with amidships of OBO ship in St Lawrence seaway.

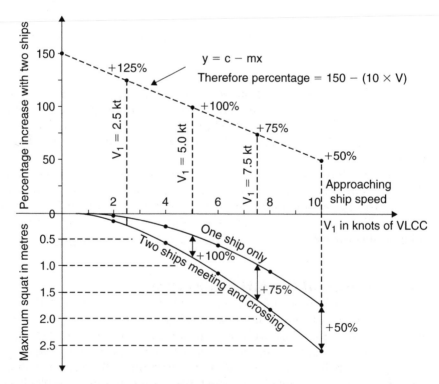

Fig. 19.4   Maximum squats for one ship, and for the same ship with another ship present.

How may these squat increases be explained? It has been shown in the chapter on Ship Squat that its value depends on the ratio of the ship's cross-section to the cross-section of the river. This is the blockage factor 'S'. The presence of a second ship meeting and crossing will of course increase the blockage factor. Consequently the squat on each ship will increase.

Maximum squat δ_{max} is calculated by using the equation:

$$\delta_{max} = \frac{C_B \times S^{0.81} \times V^{2.08}}{20} \text{ m}$$

Worked example 19.1

A Supertanker has a breadth of 50 m with a static even keel draft of 12.75 m. She is proceeding along a river of 250 m and 16 m depth of rectangular cross-section. If her speed is 5 kt and C_B is 0.825, then calculate her maximum squat when she is on the centreline of this river.

Let the blockage factor = S

$$S = \frac{b \times T}{B \times H} = \frac{50 \times 12.75}{250 \times 16} = 0.159$$

Hence
$$\delta_{max} = \frac{0.825 \times 0.159^{0.81} \times 5^{2.08}}{20}$$

$$= 0.26 \text{ m at the bow because } C_B > 0.700$$

Worked example 19.2

Assume now that this Supertanker meets an oncoming Container ship also travelling at 5 kt (see Figure 19.5). If this Container ship has a breadth of 32 m, a C_B of 0.580, a static even keel draft of 11.58 m, then calculate the maximum squats of both vessels when they are transversely in line as shown.

The blockage factor now becomes $S = \dfrac{(b_1 \times T_1) + (b_2 \times T_2)}{B \times H}$

Hence
$$S = \frac{(50 \times 12.75) + (32 \times 11.58)}{250 \times 16} = 0.252$$

For the Supertanker
$$\delta_{max} = \frac{0.825 \times 0.252^{0.81} \times 5^{2.08}}{20} \text{ m}$$

$$= 0.38 \text{ m at the bow}$$

For the Container ship
$$\delta_{max} = \frac{0.580 \times 0.252^{0.81} \times 5^{2.08}}{20} \text{ m}$$

$$= 0.26 \text{ m at the stern}$$

The maximum squat of 0.38 m for the Supertanker will be at the bow because her C_B is greater than 0.700. Maximum squat for the Container ship will be at the stern because her C_B is less than 0.700. As shown this will be 0.26 m.

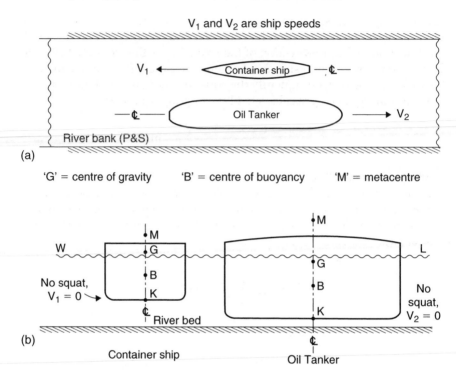

'G' = centre of gravity   'B' = centre of buoyancy   'M' = metacentre

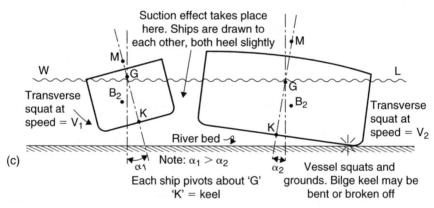

Fig. 19.5   Transverse squat, caused by ships crossing in a confined channel.

If this Container ship had travelled alone on the centreline of the river then her maximum squat at the stern would have only been 0.13 m. Thus the presence of the other vessel has doubled her squat from 0.13 to 0.26 m.

Clearly, these results show that the presence of a second ship does increase ship squat. Passing a moored vessel would also make blockage effect and squat greater. These values are not qualitative but only illustrative of this phenomena of Interaction in a ship to ground (squat) situation. Nevertheless, they are supportive of A.D. Watt's statement.

Ship to ship Interaction

Consider Figures 19.6 and 19.7, where a Tug is overtaking a large ship in a narrow river. Five cases have been considered.

Case 1   The Tug has just come up to Aft Port quarter of the ship. The domains have become in contact and Interaction occurs. The positive bulb of the ship reacts with the positive bulb of the Tug. Both vessels veer to Port side. Rate of turn is greater on the Tug. There is a definite possibility of the Tug veering off into the adjacent river bank as shown in Figure 19.6.

Case 2   The Tug is in danger of being drawn bodily towards the ship because the negative pressure (suction) bulbs have interfaced. The bigger the differences between the two deadweights of these ships the greater will be this transverse attraction. Each ship develops an angle of heel as shown. There is a danger of the ship losing a bilge keel or indeed fracture of the bilge strakes occurring. This is 'transverse squat', the loss of underkeel clearance at forward speed. Figure 19.5 shows this happening with the Tanker and the Container ship.

Case 3   The Tug is positioned at the ship's forward Port quarter. The domains have become in contact via the positive pressure bulbs (see Figure 19.6). Both vessels veer to the Starboard side. Rate of turn is greater on the Tug. There is great danger of the Tug being drawn across the path of the ship's heading and bowled over. This has actually occurred with resulting loss of life.

Case 4   The positive forward pressure bulb of the Tug has come in contact with the negative pressure bulb of the ship. Interaction has occurred (see Figure 19.7). Because the ship's pressure bulb will be the larger in negative magnitude, the Tug will pivot and be drawn towards the ship. The end result could be the Tug shearing to Starboard and hitting the Aft Port quarter of the ship's sideshell. This happened a few years ago on the River Plate in South America.

Case 5   The positive Aft pressure bulb of the Tug has come in contact with the negative pressure bulb of the ship. Interaction has occurred (see Figure 19.7). Because the ship's pressure bulb will be the larger in negative magnitude, the Tug will pivot and be drawn towards the ship. The end result could be the Tug shearing to Port and hitting the river bank.

Note how in these five cases it is the smaller vessel, be it a Tug, a pleasure craft or a local ferry involved, that ends up being the casualty!!

Figures 19.8 and 19.9 give another two cases (cases 6 and 7, respectively) of ship to ship Interaction effects in a narrow river.

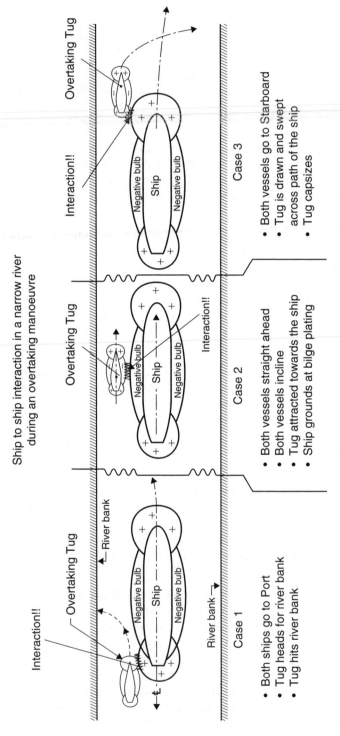

Fig. 19.6   Ship to ship Interaction in a narrow river during an overtaking manoeuvre.

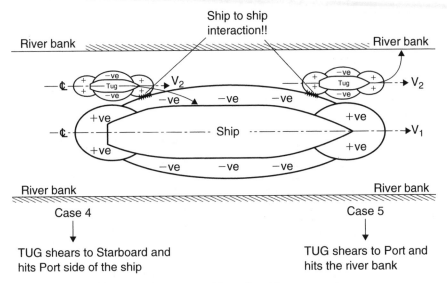

Fig. 19.7   The overtaking of a ship manoeuvre.

Methods for reducing the effects of Interaction in Cases 1–7

Reduce speed of both ships and then if safe increase speeds after the meeting crossing manoeuvre time slot has passed. Resist the temptation to go for the order 'increase revolutions.' This is because the forces involved with Interaction vary as the speed squared. However, too much of a reduction in speed produces a loss of steerage because rudder effectiveness is decreased. This is even more so in shallow waters, where the propeller rpm decreases for similar input of power in deep water. Care and vigilance are required.

Keep the distance between the vessels as large as practicable bearing in mind the remaining gaps between each shipside and nearby river bank.

Keep the vessels from entering another ship's domain, e.g. crossing in wider parts of the river.

Cross in deeper parts of the river rather than in shallow waters bearing in mind those increases in squat.

Make use of rudder helm. In Case 1, Starboard rudder helm could be requested to counteract loss of steerage. In Case 3, Port rudder helm would counteract loss of steerage. For Case 4, Port helm should be applied. For Case 5, Starboard helm would help counteract the loss of steerage.

Ship to shore Interaction

Figures 19.10 and 19.11 show ship to shore Interaction effects. Figure 19.10 shows the forward positive pressure bulb being used as a pivot to bring a ship alongside a river bank.

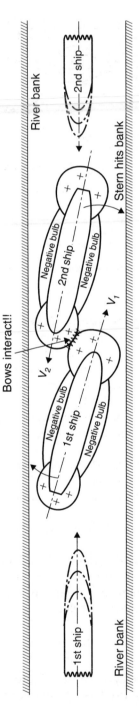

Bows interact!!

Stern hits bank

Fig. 19.8   Case 6: Ship to ship Interaction. Both sterns swing towards river banks. The 'approach' situation.

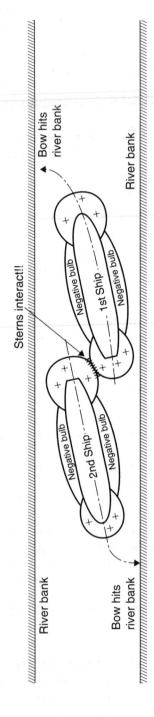

Sterns interact!!

Fig. 19.9   Case 7: Ship to ship Interaction. Both bows swing towards river banks. The 'leaving' situation.

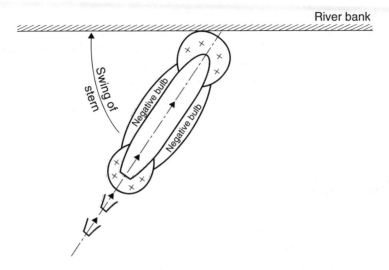

Fig. 19.10   Ship to bank Interaction. Ship approaches slowly and pivots on forward positive pressure bulb.

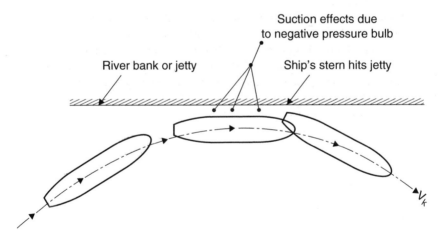

Fig. 19.11   Ship to bank Interaction. Ship comes in at too fast a speed. Interaction causes stern to swing towards river bank and then hits it.

Figure 19.11 shows how the positive and negative pressure bulbs have caused the ship to come alongside and then to veer away from the jetty. Interaction could in this case cause the stern to swing and collide with the wall of this jetty.

Summary

An understanding of the phenomenon of Interaction can avert a possible marine accident. Generally a reduction in speed is the best preventative procedure. This could prevent an incident leading to loss of sea worthiness, loss of income for the shipowner, cost of repairs, compensation claims and maybe loss of life.

Questions

1 Define the following terms:
   (a) Ship to ship Interaction.
   (b) Pressure bulbs around a moving vessel.
   (c) Ship domain.

2 List the effects of Interaction in a narrow river, when the amidships of an overtaking vessel becomes in line with the amidships of another moving vessel.

3 List *four* procedures that can be taken by shipboard personnel to reduce the effects of Interaction.

4 A small vessel has a Br. Mld of 20 m and a static even keel draft of 8 m. A larger vessel has a Br. Mld of 45 m and an even keel draft of 11.50 m. Calculate the blockage factor 'S' when these ships are in line with each other, in a rectangular canal of 231.6 m width and depth of water of 13 m.

5 Illustrate a 'ship to bank' Interaction as a ship slowly approaches a river jetty and is pivoting on a forward pressure bulb.

Chapter 20

Ship vibration

There are several terms related specifically to ship vibration, some of them being:

- Node
- Anti-node
- Mode
- Amplitude
- Frequency
- Resonance or synchronisation
- Entrained water.

Node   A node is a point in a vibrating beam or ship where the movement or amplitude is zero. In a vibrating mass there may be two, three, four or more such points. This is illustrated in Figure 20.1.

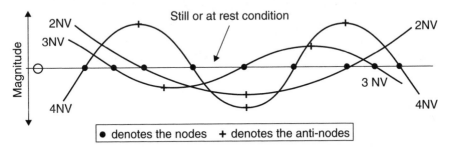

Fig. 20.1   Nodes and anti-nodes in a vibrating mass.

Anti-node   This is a point in a vibrating beam or ship where the movement or magnitude is greatest. To decrease the amount of vibration, weights can be added to the vibrating mass. It is most efficient to attach these weights at these anti-nodal points.

Mode   This is the manner or direction in which the beam or ship vibrates. As examples, vibration may be vertical (V), horizontal (H) or torsional (T).

Consequently, a vibrating mass with two nodes with vertical movement is known as a 2NV mode. A mass that is vibrating horizontally with three nodes is of 3NH mode.

Amplitude   This is the actual movement of a plate 'out to out' (see Figure 20.2). Typically it can be 3 mm. However, in F. Todd's book, 'Ship Vibration', he cites amplitude as high as 12 mm 'out to out'.

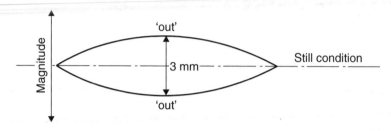

Fig. 20.2   Movement of a vibrating plate.

Frequency   This is the movement of the vibrating beam or ship measured in cycles/minute. It is the reciprocal of the time period (T), as shown in Figure 20.3.

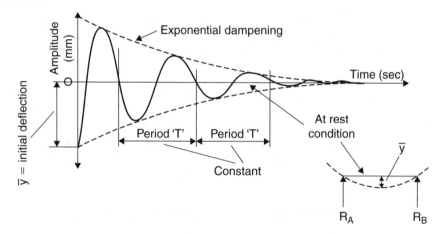

Fig. 20.3   Natural frequency 'N' where N = 1/T.

Resonance or synchronisation   This occurs when the frequencies associated with the ship are of similar value. One may be a natural frequency and one may be a mechanical frequency. Both types can be measured in cycles/minute.

When frequencies are of the same value (synchronous) then vibration problems become of concern to those on board ship (see Figure 20.4). This is because the magnitudes are exacerbated.

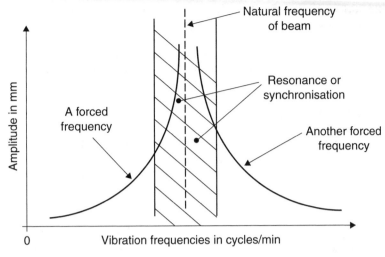

Fig. 20.4   Natural and forced frequencies, with resonance.

Entrained water   This is the amount of water picked up by the ship as she moves ahead at speed. The amount picked up and set in motion will depend upon the surface condition of the hull, the ship's breadth and the ship's draft. In effect, it is like a 'waistcoat' that the vessel carries along with her. (see Figure 20.5 and Worked example 20.2).

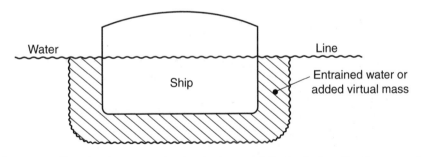

Fig. 20.5   Entrained water as 'waistcoat' around the hull of a vibrating vessel.

Table 20.1 shows feedback from several ships for various nodes, modes and frequencies. The frequencies are all in cycles/minute.

After estimating the natural hull frequency (N), it is possible to show how it changes for modification to the C_B, modification to the ship's draft or modification to the ratio of water depth (H)/ship's mean draft (T) value.

Figure 20.6 illustrates the effect on the frequency of changing the C_B from 0.550 to say 0.850.

Table 20.1   Measured natural hull frequencies for several ships (note the very small frequency for the VLCC*)

Type of ship	LBP (m)	Displacement (tonnes)	2NV mode	2NH mode	3NV mode	4NV mode
Oil Tanker	147	19 120	74	110	185	–
Cargo-passenger ship	143	16 358	85	116	155	221
Passenger-cargo ship	127	8600	123	180	237	315
Bulk Ore carrier	160	16 200	45	120	106	168
VLCC	300	250 000	35	35	–	–

* Very Large Crude Carriers.

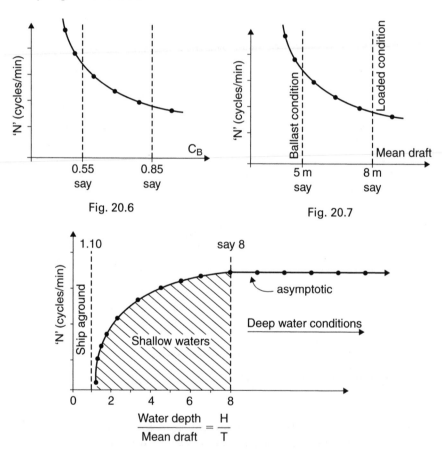

Fig. 20.6

Fig. 20.7

Fig. 20.6–20.8   Changes in 'N' for changes in C_B, mean draft and H/T values.

Figure 20.7 illustrates the effect on the frequency of changing the mean draft from 5 to say 8 m.

Figure 20.8 illustrates the effect on the frequency of changing the H/T ratio from 1.10 to say 8.00.

Causes of vibration

1. One of the main causes of vibration on ships is due to unbalanced forces where reciprocating machinery is fitted.
2. Forces may be present due to forcing impulses in an Internal Combustion engine. Diesel engines can cause problems due to their required rpm being close to the natural hull frequency. Steam Turbine machinery, due to having no unbalanced parts, cause little or no vibration problems of this type.
3. Most vibration problems on ships can be traced to the propeller. This may be because of the following:
   (a) Irregular flow towards the propeller disc area.
   (b) Damaged propeller. It could be a broken blade. It could be a bent blade.
   (c) Unbalanced new propeller – unlikely as the propeller manufacturers will endeavour to produce one that is perfectly balanced. As a matter of pride, the centre of mass of the propeller will, in all probability, be at the centre point of the propeller disc.
   (d) Propeller is too large for the aperture adjacent to the sternframe and the rudder. That is to say, the propeller clearances are too small (see Figure 20.9).
   (e) Pitch variation from propeller root to propeller tip varying slightly from blade to blade.
   (f) Propeller has wrong number of blades and so produces resonance with another frequency linked with the vessel. This is numerically shown later with a worked example.
4. Sea effects.
   A ship will vibrate due to the pounding effect at the forward end and also due to wave frequency acting at the same frequency as that of the hull. This type of vibration is known as 'whipping' and is experienced mostly on fast Container ships.

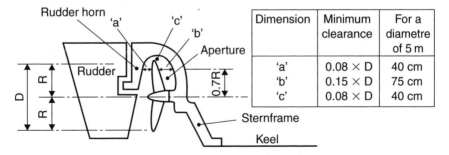

Fig. 20.9   Propeller clearances around a marines rudder arrangement.

Reduction of vibration on ships already built

The source of vibration can be located by using a vibrograph. This measures the amplitudes at the resonant frequency.

If it is an auxiliary unit that is giving vibration problems, the following procedures can be considered and implemented:

(a) Alter the rpm to avoid synchronisation.
(b) Improve any out of balance components within the unit.
(c) Fit elastic seating. This can be in the form of hardened rubber mountings or metal springs. These will dampen movement, like pads at the base of a lathe or shock absorbers on a motorcycle.

If the main machinery is causing vibration problems then consider the following:

(a) Alter the engine rpm to avoid synchronisation.
(b) Add balance weights to decrease or eliminate out of balance forces.
(c) If she is a twin-screw ship, the rpm on the Port side could be adjusted so that the propellers are turning at slightly different revolutions. Nowadays, this can be done using electronic sensors.

If the vibration problem is local, say in a cabin, small stiffeners can be welded into place. This can reduce the amplitudes, e.g. in the deck plating. This solution is not too efficient because it could move the problem along the deck and into the next cabin. That next cabin may be your cabin!! Also with this procedure a lot of extra steel may have to be fitted to solve the vibration problem.

If the problem is traced to the propeller, then consider the following:

(a) Fit a new propeller, fully balanced, with no bent or broken blades.
(b) Fit a new propeller with a different number of blades, e.g. four blades instead of three blades (see later Worked example 20.1).
(c) Increase the propeller clearances by raking the propeller blades in an Aft direction (see Figure 20.10).
(d) Fill the Aft Peak tank with water ballast. This greatly helps dampen down forces emanating from the propeller, being transmitted upwards towards the accommodation and navigation spaces.
(e) Alter the loading of the ship. Add or discharge water ballast. This will change the natural hull frequency and perhaps move it to a frequency that is not resonant with the propeller or engine revolutions.
(f) Fit a Kort nozzle or Kort rudder around the propeller. If filled with polyurethane foam, the propeller forces will be further dampened and the vibration problems decreased.

If the vibration problems are caused by sea effects, simply consider the following:

(a) Alter the course heading of the ship. This could help avoid resonance with another frequency like the vessel.
(b) Increase or decrease ship speed to avoid wave excited vibration.

On ships, vibration is similar to corrosion in that it will always exist in some form or other. However, by careful planning, they can usually be reduced to a level acceptable for those on board ship.

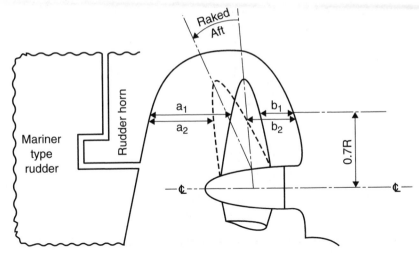

Fig. 20.10   Propeller blades raked Aft in order to increase
aperture clearance b_1 to b_2.

Which propeller to fit?

Worked example 20.1

A Passenger–cargo ship has the following natural hull frequencies (measured in cycles/min):

$$2NV = 123 \quad 2NH = 180 \quad 3NV = 237 \quad 4NV = 315$$

This vessel is to be fitted with a *three*-bladed propeller or a *four*-bladed propeller each operating at 105 rpm. In order to reduce vibration problems, determine which is the better propeller to fit to this vessel:

For the *three*-bladed propeller, the 1st harmonics = 123, 180, 237 and 315

For the *three*-bladed propeller, the 2nd harmonics = 123/3, 180/3, 237/3
and 315/3

= 41, 60, 79 and 105

Note how the 2nd harmonics are the 1st harmonics divided by the number of blades fitted on the propeller:

For the *four*-bladed propeller, the 1st harmonics = 123, 180, 237 and 315

For the *four*-bladed propeller, the 2nd harmonics = 123/4, 180/4, 237/4
and 315/4

= 30.75, 45, 59.25 and 78.75

Figure 20.11 shows that the propeller revolutions of 105 would be synchronous with the highest value of the 2nd harmonics. Consequently, severe and

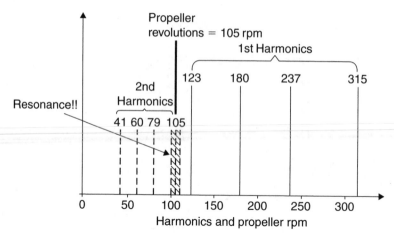

Fig. 20.11   1st and 2nd harmonics for a three-bladed propeller. At resonance, forcing frequency equals natural frequency.

uncomfortable vibration would exist as the natural frequency and the forcing frequency are resonant. Suggest changing this three-bladed propeller for a four-bladed propeller and review the situation.

Figure 20.12 shows that the four-bladed propeller would be ideal. This is because the 105 propeller revolutions are well away from an area of resonance with any of the 1st or 2nd harmonics. It is 18 cycles/min below the lowest 1st harmonic of 123 and 26.25 above the highest 2nd harmonic of 78.75 cycles/min.

Conclusion: Fit the *four*-bladed propeller instead of the *three*-bladed propeller.

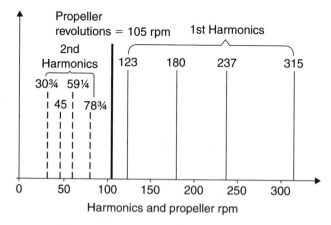

Fig. 20.12   1st and 2nd harmonics for a four-bladed propeller. No resonance now with propeller rpm.

Ship vibration frequency calculations

Worked example 20.2

An Oil Tanker has the following vibration particulars:

LBP = 147.2 m, displacement = 20 327 tonnes, Draft Mld = 8.42 m, I_{NA} = 43 m^4.
Br. Mld = 20 m, Schlick's Ø value = 2.7×10^6, Todd's β value = 108 450. Depth
Mld = 11.35 m, Burrill's 'r_s' factor = 0.132.

Estimate the 2NV natural hull frequency 'N' in cycles/minute using the research work of: (a) Otto Schlick, (b) F. Todd, (c) F. Todd and W.J. Marwood and (d) Prof L.C. Burrill.

(a) Schlick's method

$$N = Ø \times \left[\frac{I_{NA}}{W \times L^3}\right]^{0.5} \text{cycles/min} = 2.7 \times 10^6 \times \left[\frac{43}{20\,327 \times 147.2^3}\right]^{0.5}$$

Thus N = 69.57 cycles/min

(b) Todd's method

$$N = β \times \left[\frac{B \times D^3}{W \times L^3}\right]^{0.5} \text{cycles/min}$$

This is a step forward because, being based on main dimensions, it does not require I_{NA} and so does not need a preliminary midship section scantlings plan for the new ship:

$$N = 108\,450 \times \left[\frac{20 \times 11.35^3}{20\,327 \times 147.2^3}\right]^{0.5}$$

Thus N = 72.83 cycles/min

(c) Todd and Marwood's method

Let the total virtual mass = W_2

$$W_2 = W \times \left[\frac{B}{3 \times d} + 1.2\right] \text{tonnes}$$

$$= 20\,327 \times \left[\frac{20}{3 \times 8.42} + 1.2\right] = 40\,487 \text{ tonnes}$$

This represents the total vibrating mass. It is made up of the ship's displacement plus the entrained water surrounding the vessel. As can be observed this mass W_2 is almost *twice* the ship's displacement:

$$N = 2.39 \times 10^6 \times \left[\frac{I_{NA}}{W_2 \times L^3}\right]^{0.5} + 28 \text{ cycles/min}$$

Thus $N = 2.39 \times 10^6 \times \left[\dfrac{43}{40\,487 \times 147.2^3} \right]^{0.5} + 28$

$N = 43.61 + 28 = 71.61$ cycles/min

(d) Burrill's method

$$N = A_1 \times A_2 / A_3$$

where:

$A_1 = 4.34 \times 10^6$ and is Burrill's vibration coefficient

$A_2 = \left[\dfrac{I_{NA}}{W \times L^3} \right]^{0.5}$ and is identical to Schlick's collection of variables under the root sign

$A_3 = \left[\dfrac{1 + B}{2 \times d} \times (1 + r_s) \right]^{0.5}$

The first factor in A_3 is a virtual inertia factor, along the lines suggested by Todd and Marwood. The second factor in A_3 is a shear correction factor suggested by Burrill:

So $N = \dfrac{(4.34 \times 10^6) \times [I_{NA}/(W \times L^3)]^{0.5}}{\{[(1 + B)/(2 \times d)] \times (1 + r_s)\}^{0.5}}$

$A_2 = \left[\dfrac{43}{20\,327 \times 147.2^3} \right]^{0.5}$ $A_3 = \left[\dfrac{1 + 20}{2 \times 8.42} \times (1 + 0.132) \right]^{0.5}$

Thus $N = 1.01$ cycles/min

Some vibration approximations

Schlick's coefficient $\emptyset$ may be approximated, using:

Schlick's $\emptyset = 3.15 \times 10^6 \times C_B^{0.5}$ C.B. Barrass (1992)

F. Todd's β value may be approximated, using:

F. Todd's $\beta = 124\,000 \times C_B^{0.6}$ C.B. Barrass (1992)

$2NH/2NV = 1.33$ approximately   General Cargo ships

$2NH/2NV = 1.50$ approximately   Oil Tankers

$3NV/2NV = 2.50$ approximately   Oil Tankers

Conclusions for Worked example 20.2

Schlick's formula and method is perhaps the weakest because it was suggested way back in 1894. Proportions, size and style of ship designs have changed greatly over the intervening years, so his coefficient Ø must be treated with caution.

The average of the other three methods gives a value for the 2NV mode of 71.7 cycles/min for the natural hull frequency.

All these coefficients must be carefully used for making predictions of hull frequencies. Comparisons must only be made between ships that are very similar in type, size and type of structural steel design. As time goes by these coefficients will gradually change as new ship types and design concepts evolve.

They cannot be used for Supertankers. This is because of the tremendous dimensions of these ships. The 2NV hull frequency can be different at Upper Deck level when compared to that measured at bottom shell level.

Questions

1 A new vessel has a C_B of 0.745. Estimate the vibration coefficients suggested by O. Schlick and F. Todd.

2 Using a diagram, show how the amplitude of a simply supported beam decreases with time.

3 Show graphically how the natural hull frequency (N) decreases as a ship changes from ballast condition of loading to fully loaded condition.

4 List *five* methods of reducing vibration problems in existing ships.

5 An Oil Tanker has the following particulars:
   LBP = 154 m,   displacement = 22 200   tonnes,   Draft   Mld = 8.40 m,
   I_{NA} = 59 m⁴, Br.Mld = 21.4 m.
   Use the Todd and Marwood method to estimate the natural hull frequency for the 2NV mode. Proceed to approximate the 2NH mode for this ship.

6

Vessel	LBP (m)	W (tonnes)	2NV (cycles/min)	I_{NA} (m⁴)
Basic ship	120	12 500	76.15	18.00
New similar design	125	14 800	xx.xx	20.85

Estimate the 2NV natural hull frequency for the new similar design based on the data given in the above table.

7 (a) What are 2nd harmonics?
   (b) How are 2nd harmonics calculated?
   (c) Why are 2nd harmonics calculated?

Chapter 21

Performance enhancement in ship-handling mechanisms

Constant research and development are being made to improve the performance of ships. The field of research may be in:

- Increased service speed for similar input of engine power.
- Lower oil fuel consumption per day for similar service speed.
- Better manoeuvrability in confined waters.
- Anti-rolling devices.
- Alternative methods of propulsion.
- Stopping of ships.
- Reduction of frictional resistance.
- Reduction of wave-making resistance.
- Better onboard maintenance management programmes.
- Reduction of time spent on loading and discharging in Port.
- More use of non-conventional materials to replace mild steel.

Ship-handling mechanisms

Becker twisted rudder
Designed to greatly reduce rudder cavitation and to improve the manoeuvrability performance of a full spade rudder. When compared to conventional rudders, it gives greater cavitational-free rudder zones. Maintenance costs are reduced. Improvements in acoustic performance are achieved. This design of rudder is fitted on the 25 kt 'MV Seafrance Rodin' (see Figure 21.1).

Schilling rudders
These rudders can go to 70° to Port or Starboard without stalling (see Figure 21.2). They have low cost and maintenance. The unit is a one-piece balanced rudder that can turn a ship within her own length.

Schilling VecTwin rudders
With this concept, two rudders operate independently behind a single propeller (see Figure 21.3). This allows full control of the propeller thrust. Best fitted on coastal and river craft.

Fig. 21.1   Profile of a twisted rudder.

Fig. 21.2   A Schilling rudder.

Activated stabilising tanks

These are two tanks Port and Starboard (P&S) situated at or near to amidships (see Figure 21.4). When the ship is upright each of these tanks is half full of water. As the ship rolls, water is mechanically pumped vertically up

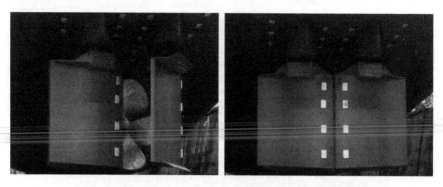

Fig. 21.3 The Becker Schilling 'VecTwin' rudders.

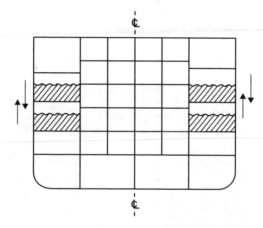

Fig. 21.4 Activated stabilising tank. Situated at the amidship.

or down. This creates an anti-rolling moment thus creating a dampening effect on the existing rolling motions.

These tanks are particularly efficient on Container vessels and have helped considerably reduce cargo damaging the sides of the containers.

Figure 21.5 illustrates another method where water is pumped across the ship with the aid of a high-powered pump. Water is quickly transferred across the ship at high pressure between two wing tanks P&S as shown in the sketch.

Tee-duct in Fore Peak Tank

This duct can be used for turning or stopping the ship (see Figure 21.6). For a 165 000-tonne Very Large Crude Carrier (VLCC), British Ship Research Association (BSRA) estimated a saving in the distance of about 30% of the crash-stop value.

Brake flaps

Brake flaps have been fitted on a 165 000-tonne VLCC. BSRA have established that it would require a hinged flap of 7.60 m depth by 4.88 m set up

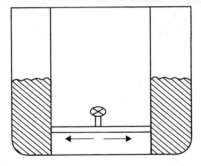

Fig. 21.5   Activated stabilising tank. Situated near amidships.

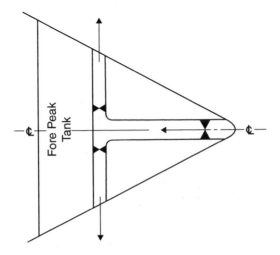

Fig. 21.6   Tee-duct in Fore Peak Tank.

to an angle of 60° (see Figure 21.7). The predicted saving in stopping distance is then about 20% of the crash-stop value. Brake flaps may also be telescopic as shown in Figure 21.8.

Submerged parachutes

These can be up to 14 m in diameter. One is placed to P&S, at about amidships (see Figure 21.9). When the ship's speed has decreased to about 7 kt they open in the water. BSRA have suggested a saving in the stopping distance of about 30% of the crash-stop value.

Stern fins

Also known as Grouthues–Spork spoilers (see Figure 21.10). These are fins or strips of steel welded around the sternframe of a ship. The idea is that these welded fins will direct water into the propeller disc and thus improve the efficiency of the propeller. Cheap to fit and retrofit. Claims of up to 6% power savings have been made for full-form and medium-form hulls.

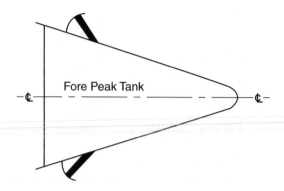

Fig. 21.7   Brake flaps.

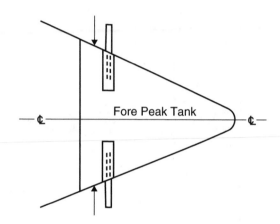

Fig. 21.8   Brake flaps can be telescopic.

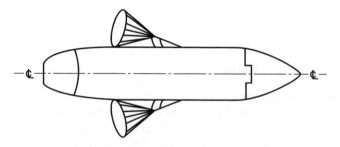

Fig. 21.9   Submerged parachutes.

Hinged tail flap in rudder

This design of rudder can cause the turning circle diameter to be halved (see Figure 21.11). Best fitted on riverboats, supply vessels, research ships and fishing vessels. On some designs, the hinged flap can be removed to increase the course-keeping properties. However, the turning characteristics of the ship would then deteriorate.

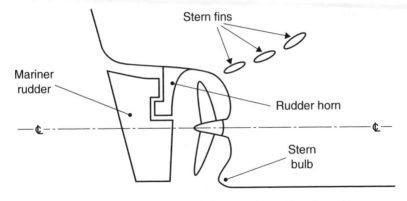

Fig. 21.10   Stern fins (aerofoils). Grouthues–Spork spoilers.

Fig. 21.11   Becker tail-flap rudder.

Kort nozzle

This is a *fixed* cylinder of plating fitted around the propeller. The wall of the cylinder is streamlined and is often fitted with polyurethane foam to dampen vibration from propeller rpm. The nozzle also decreases the pitching of ships in heavy seas.

Blade clearances are usually ½% of propeller diameter, although some designs with different blade characteristics have shown clearances as high as 30 cm. The length of nozzle for'd and aft is about half the propeller diameter.

Nozzles have been fitted on tugs, ferries, trawlers and VLCCs. On tugs and ferries the diameter is about 2–3 m. On some VLCCs, these fixed nozzles can be as much as 8.0 m in diameter.

For the same input of engine power, with a fixed nozzle, it is claimed that there is 20% extra thrust or about 8% increase in ship speed. The most recent designs have the propeller *aft* of the Kort nozzle and not actually inside it. On smaller ships, this fixed nozzle can have an aft raked axis up towards the stern of 5–7° as shown in Figure 21.24.

Kort rudder

This is similar to the nozzle except that it is *not fixed* (see Figures 21.12 and 21.14. It can rotate to about 35° P&S. The advantages are similar, plus the

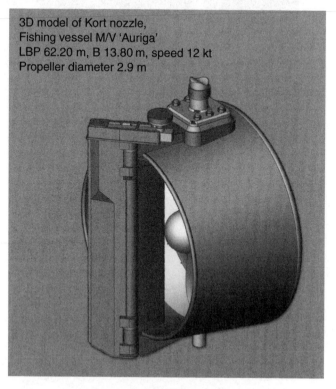

3D model of Kort nozzle,
Fishing vessel M/V 'Auriga'
LBP 62.20 m, B 13.80 m, speed 12 kt
Propeller diameter 2.9 m

Fig. 21.12   Becker Kort steering nozzle.

advantage of increased manoeuvrability when going astern. There is no need to have a conventional rudder fitted.

The Kort rudder is only fitted on smaller ships such as tugs. It is found to be very effective on these vessels. Some trawlers have been fitted with a Kort rudder P&S. This rudder gives superior astern manoeuvrability and reduced turning circles. Figure 21.13 shows a vessel fitted with twin nozzles.

Fig. 21.13   Vessel with Twin Kort steering nozzles and Becker flaps.

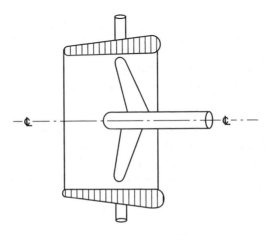

Fig. 21.14   Kort rudder.

Bulbous bows

The main advantage is the gain in speed for similar input of engine power. This is because the bulb assists in reducing the wave-making resistance (see Figure 21.15). At loaded condition, this increase in speed is ¼–½kt. In ballast condition the speed increase can be ½–¾kt. Another advantage of fitting bulbous bows is that they give added strength in the For'd Peak Tank. Also, they help reduce vibration for'd of the collision bulkhead.

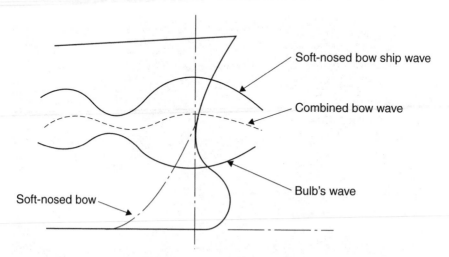

Fig. 21.15   Bulbous bows.

However, they are more costly to build than the conventional soft-nosed bow. Best fitted on ships with high kinetic energy ($\frac{1}{2}MV^2$). Consequently, they are best fitted on Supertankers (high mass) and Container ships (high service speeds).

Note of caution: On small vessels (less than 4000 tonnes dwt) with low service speeds (less than 12 kt) eddy making occurs at the bulb. This leads to drag and increases in resistance.

Some vessels have a bulb form fitted at the stern, again to cut down on the ship's wave-making resistance (see Figure 21.10).

Rotating cylinder rudders

This design has a rotating cylinder on the for'd edge of the rudder (see Figure 21.16). When in operation the rudder may be turned to 90° P or S. One vessel fitted with this design of rudder is the 14 300-tonne RO-RO vessel *Rabenfels*. She is owned by the Hansa Shipping line. The cylinder, on this 195 m length between perpendiculars (LBP) ship is 3 m long and 0.50 m diameter. It rotates at 300 rpm when rudder helm is 75° P&S.

This ship is also fitted with the transverse thrusters. National Physical Laboratory (NPL) have predicted that for a 250 000-tonne VLCC, the rotor

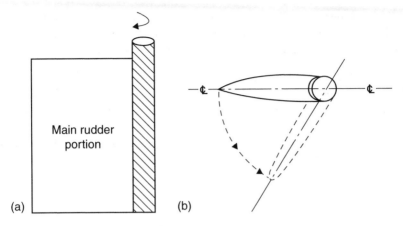

Fig. 21.16   Rotating cylinder on the for'd edge of the rudder.

would be about 60 cm diameter. Equivalent bollard pull is about 130 tonnes. The cylinder can rotate in anti-clockwise and clockwise directions. The vessel turns in her own length.

Hydraulic fin stabilisers

These fins assist in reducing the transverse roll of ships (see Figure 21.17). A roll of 30° P&S with these stabilisers inboard may exist. This can be changed

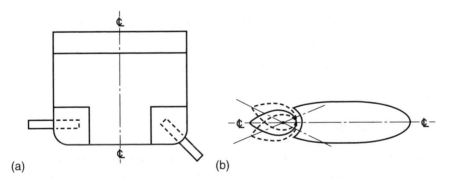

Fig. 21.17   Hydraulic fin stabilisers.

to 1.5° P&S with these stabilisers fully extended. Rather expensive to maintain and repair. Fitted mainly on Passenger Liners such as the QE11.

These fins may hinge into the sides of a vessel or move outwards in horizontal movement. Most fins move out horizontally at or very near to amidships. Some fins extend outwards P&S at an angle of 45° through the bilge plating.

The fins may be built in two parts. One will be a fixed vane. The other will be a moving vane. This moving vane assists in producing a dampening effect on the rolling motions of a ship in a heavy seaway.

Twin-hull ships

These are Small Waterplane Area (WPA), Twin-Hull ships or Ships With A Twin Hull, known as SWATH vessels (see Figure 21.18). The SWATH has a two separate hull forms joined together by a horizontal transverse steel bridge-type construction. SWATH vessels are very suitable as cross-channel ferries in reasonably calm waters.

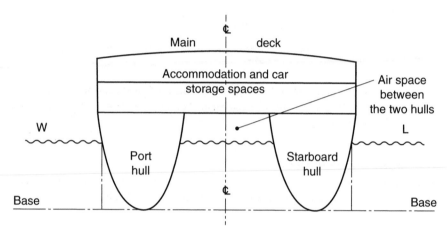

Fig. 21.18   The SWATH design (ship with a twin hull).

When at high speed they plane and ride up on the waves. However, SWATH vessels pitch more than mono-hull vessels and are not safe in bad weather conditions.

Double-skin hulls

Now being fitted beneath and at the side of the main Cargo network of tanks on all new Supertankers (Figure 21.19). The main reason is because of a number of bad accidents causing oil spillage slicks and harming the environment. It is hoped that by fitting these double skins, it will lead to damage limitation in the event of a collision or grounding.

John Crane Lips rudders

Formerly known as the Wärsillä Propac rudder (see Figure 21.20). Fuel consumption and power reductions in the order of 5–6% are predicted, together with a 25% reduction in vibration levels.

This design has been fitted on the 'Finnmarken', a Passenger vessel having two four-bladed controllable pitch propellers. This concept has rudder horns, flaps and fixed torpedo bulbs that are streamlined into the propellers.

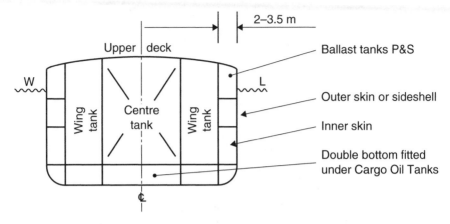

Fig. 21.19   The double-skin Oil Tanker design.

The two CP propellers in position at Kleven Verft on the new Hurtigruten ship *Finnmarken*, showing the John Crane Lips efficiency rudders (formerly known as the Wärtsilä Propac). Fuel consumption and power reductions in the order of 5–6% are predicted, together with a 25% reduction in vibration levels. The Ulstein swing-up azimthing thruster can also be seen in this illustration in its stowed position (right).

Fig. 21.20   John Crane Lips efficiency rudders.

Rudder fins

After being in service for a while some ship rudders suffered from cavitation at their lower regions. Kawasaki suggested a solution for this in that they welded four fins in an 'X' formation around a bulb built into the rudder. Known as rudder bulb-fins (see Figure 21.21).

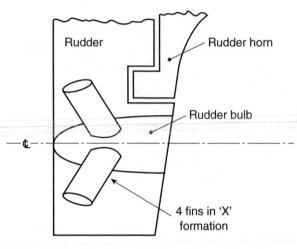

Fig. 21.21   Rudder bulb-fins. Kawasaki design.

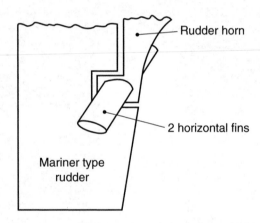

Fig. 21.22   Additional thrusting-fins. Ishikawajima–Hirama design.

Ishikawajima–Hirama had a similar idea of fitting fins. They fitted two horizontal fins at the base of the rudder horn (see Figure 21.22). Like the Kawasaki fins, these additional fins reduced the cavitation at the lower regions of the rudder.

Stern tunnel

It is like a railway tunnel built over the propeller helping to ensure that the propeller is kept sufficiently immersed when vessel is at ballast drafts (see Figure 21.23). Good for dampening vibration forces, especially where the propeller diameter is large and operating with lower than usual rpm.

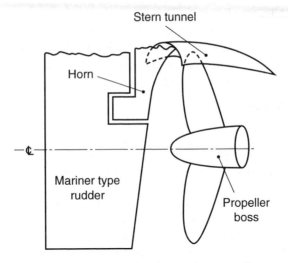

Fig. 21.23   Stern tunnel over the propeller.

Activated rudder

This is a more robust design of rudder and sternframe to dampen down rolling of small ships in heavy seas. When a vessel rolls to Starboard, the rudder automatically moves to Starboard. This creates an anti-rolling couple, causing the ship to return to the upright condition (See Figure 21.24).

If the ship rolls to Port, then rudder moves to Port. If ship is upright, then the rudder automatically returns to zero helm. NPL have tested this rudder on a trawler. From their subsequent film have shown the design to be very efficient on this size and type of ship.

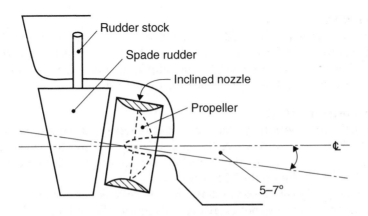

Fig. 21.24   Inclined nozzle with a spade rudder.

Pleuger rudder

This is an active rudder with a rotor in the aft end of the rudder. This design can be operated at rudder helm of 70° P&S. Fitted on over 100 ships, it is most efficient when the ship's speed is zero or very nearly zero. It has now being superseded by transverse thrusters.

Hull form of ship

If the width of water ahead of a ship is adequate, the rudder helm is used to turn the ship first to Port and then to Starboard, to Port and finally to Starboard. For 60% of the time taken on this manoeuvre, the for'd rpm of the propellers are used, and then the astern power is finally applied over the remaining time.

By using rudder cycling to stop the ship in this manner, the stopping time can be reduced to 80% of the stopping time of a Crash-stop manoeuvre. To be able to carry out this manoeuvre, the rudder and rudder-stock obviously must be of robust and rigorous construction.

Engine type

If a ship is fitted with Diesel machinery, the ship will stop in about 70% of the stopping distance and about 70% of the stopping time, compared to if she was fitted with Steam Turbine machinery during a Crash-stop manoeuvre. This is because, for Diesel machinery, the astern power is about 80% of the ahead power. For Steam Turbine machinery, it is only about 40%.

Hull surface polymer paints

In order to reduce the frictional resistance to a minimum value much work has been undertaken on anti-fouling paints. These paints help counteract animal and vegetable growth on the hull of ships. This growth can be as thick as 60 mm on some vessels in various parts of the world.

Self-polishing paints are helping to reduce the roughness of hulls so much so that dry-dockings every 12 months can be extended (via dispensation) to 18–24-month intervals.

Planned maintenance

The modern idea for planned maintenance is to have a lower number of seafaring crew and for a shore-based crew to visit, inspect and carry out any maintenance required whilst the ship is in Port. This means fewer cabins required and fewer stores needed for new ships.

Asymmetrical stern

This is a German design by Nonnecke. It is aimed at reducing separation in the after-body of a ship where the flow is influenced by the clockwise or anti-clockwise rotation of the propeller. The quasi-propulsive coefficient (QPC) is increased with this type of stern.

Obviously, this asymmetrical stern can only be considered very early in the lines planning stage and not as an idea for a retrofit. From 1982 to 1987, some 30 vessels were built or in the process of construction with this type of stern.

Retrofits

Many ships have had their designed structures altered *after* being in service for a while. This alteration is known as a retrofit. It is hoped that a retrofit improves the day-to-day performance of a ship or possibly leading to a reduction in operational costs. Retrofit costs must be carefully balanced against financial gains for the shipowner.

The 'Esso Northumbria', a Supertanker, suffered with bad vibration problems in the first months of her life at sea. She was brought back to her shipbuilders for alterations. Several tonnes of steel were added to stiffen up the vessel. This retrofitted extra steel in fact reduced the vibration problems to an acceptable level.

There are cases of ship undergoing 'ship surgery' by cutting the ship at mid-length and welding in an additional length of midship section. As well as increasing the deadweight for the shipowner, the vessel also ends up (at similar input of engine power) with a slight increase in speed. This is due to the higher LBP/Breadth Mld making her a better hydrodynamic hull form. Increased deadweight will lead to increased income for the shipowner.

The 'Lagan Viking', a RO-RO vessel operating between Liverpool and Belfast actually had the depth of her holds increased. This was done by horizontally cutting right along the ship's length and raising the deck. It then meant in the greater hold depth, lorries could be stowed instead of just cars. The 'Merchant Venture', another RO-RO vessel also had this type of retrofit in order to transport trailers.

The Japanese have even gone one step better. They took a ship and cut along the centreline the entire length from bow to stern. They pulled the P&S sides apart and created a gap. Finally, they welded a new steel structure into this gap, extending from the bow to stern. As well as giving increased deadweight, the transverse stability of the vessel was improved.

Other examples of retrofits are:

- Replacing three-bladed propellers with four-bladed propellers to avoid resonance.
- Replacing Steam Turbine machinery with Diesel machinery to reduce the daily fuel consumption. Mobil Shipping Company did this with several of their Supertankers.
- Replacing a bulbous bow with a soft-nosed conventional bow to reduce resistance. This unusual retrofit was completed on a vessel called the 'Pioneer' built in Spain and owned by Manchester Liners Ltd. The conventional bow did, as predicted, give an increase in speed for a similar input of engine power.
- Computer packages with text, values and graphics, replacing: the old 'Ralston' trim and stability evaluator, the old 'Kelvin Hughes' mechanical trim and stress indicator, the electrical 'Loadicator' that gave shear forces and bending moments along the ship's length.

These last three shipboard machines are now almost obsolete.

Chapter 22

Improvements in propeller performance

Voith-Schneider Propulsion unit

This Voith-Schneider Propulsion (VSP) unit has blades that are vertically suspended and rotated about a vertical axis (see Figure 22.1). It can be moved to produce thrust in any required direction in 360°. Number of vertical blades may be four, five or six. The diameter of these propulsion units range from about 1.20 m up to a maximum of 4.40 m. The maximum depth of the blades is about 2.7 m.

These units are fitted on tugs (known as water tractors), double-ended ferries, passenger ships, buoy tenders, floating cranes, mine-hunters and oil skimming vessels. Vessels may be fitted with one, two, three or four of these units. The longest vessel fitted with this unit is a drill-ship of 122 m length and a service speed of 13 kt.

The advantages of these VSP units are:

- Propulsion and steering of the ship are integrated in a single system.
- Fast stopping with precise manoeuvring and dynamic positioning.
- A proven performance on the highest possible safety level for passengers, cargo and for the ship itself.
- The highest possible availability of ship to fulfil the transport function with the highest possible reliability to keep to the time schedule.
- An ease of maintenance, to keep the repair and overhaul intervals as short as possible.
- A simple and robust system.
- A minimum of complex control systems.
- They are environmentally friendly with low wash and noise levels.

The bollard pull on tugs for these units ranges from 25 up to 95 tonnes. The unit dispenses with the need for a conventional rudder and can turn a ship in her own length. Protection plates are fitted on the base of the propulsion unit, for when the ship has to go into dry-dock.

(a)

(b)

Fig. 22.1   VSP units. Blade depth indicated by height of man. Bottom of shell
plating is shown in (b).

Transverse thrusters

These are most efficient when the forward speed of ship is *zero*. They do not function efficiently if the ahead speed is greater than 3 or 4 kt. One diameter of duct is that of 3.3 m diameter, fitted by KaMeWa. Any larger diameter would lead to serious decreases of strength in the Fore Peak Tank (see Figure 22.2).

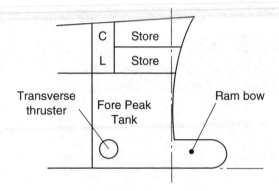

Fig. 22.2   Transverse thruster.

They may be fitted forward or aft. The 'Liverpool Bay' (a container vessel) had *two* transverse thrusters forward and one aft. As well as turning a vessel, these thrusters can be used to move a ship parallel and bodily away from a jetty into mid-river. They help reduce the call (and cost) for tugs in confined waters.

Schottel thrusters

These units have a telescopic design (see Figure 22.3). The unit descends downwards through the bottom shell to produce thrust through 360°. On the *Forties Kiwi* (160 m length overall (LOA) and 16 435-tonne dwt), four Schottel thrusters were fitted.

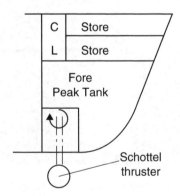

Fig. 22.3   Schottel thruster.

Grim vane wheel

This is a freely revolving wheel having a 20% larger diameter than that of a driven propeller in whose wake it is placed so that it can also reclaim some tip vortex energy (see Figure 22.4). The Grim wheel turns in the same direction as the driven propeller. Feedback from vessels indicates fracture of the wheel itself. Perhaps more research and development are required to give the idea more support within the Marine Industry. However, it has possibilities.

Fig. 22.4   The Grim vane wheel by Prof. O. Grim (ca. 1980–1983).

Groningen Propeller Technology propeller

This tip-plate propeller has been fitted on the 'Aquatique', a 3200-tonne dwt multi-purpose cargo coaster in March 2001. GPT blades exploit double-tip plates (offset on either side of each blade).

It is different from competitive designs in that it features a double tip (one on the suction side and one on the pressure side) that can be symmetrical or offset. The 'Aquatique' has a propeller diameter of 2.8 m, four-bladed controllable-pitch design with a service speed of 12.5 kt (see Figure 22.5).

A gain in ship speed (for similar input of power) equating to 0.33–0.35 kt has been recorded for this GPT equipped ship. For a similar speed, the measured power reduction amounted to 12–14%. Shipboard observers noted that noise and vibration were significantly less.

At the end of 2001, this design was retrofitted to the 'MT Nordamerika' a 35 000-tonne dwt products tanker. The propeller on this ship is 5.8 m diameter, four-bladed fixed-pitch design. Ship Trials were first carried out on with a conventional propeller. After their completion, the ship was then dry-docked and fitted with a Kappel propeller and sent back on trials (see Figure 22.6).

At the 15-kt condition the vessel's power requirement was reduced by approximately 4% (see Figure 22.7). This will result in a corresponding

(a) An important milestone for Groningen Propeller Technology: a tip plate propeller ready for application to *Aquatique*, a 3200-tonne dwt multi-purpose coaster, to test fuel saving and noise reduction. Sea trials gave superior results over conventional CP propellers or sister ships

(b) The new propeller fitted to the Saimax-class Cargo Ship *Aquatique*

Fig. 22.5   The GPT tip-plate propeller design.

Fig. 22.6   The Kappel propeller.

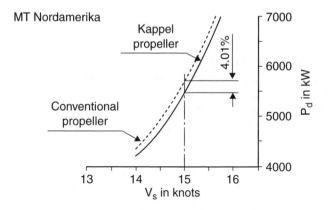

Fig. 22.7   Improved performance with the Kappel propeller.

reduction in fuel consumption, fuel costs and exhaust emissions. Stone Manganese Marine Ltd of Birkenhead (UK), a partner in the design consortium, observed that cavitational phenomena showed a marked improvement and that propeller excitation forces were similarly much improved.

Propeller Boss Cap Fins

These units are saving energy on over 830 ships around the world (see Figure 22.8). Improves propeller efficiency. By recovering the energy of the hub vortex, the Propeller Boss Cap Fins (PBCF) produces the following effects, with no maintenance costs.

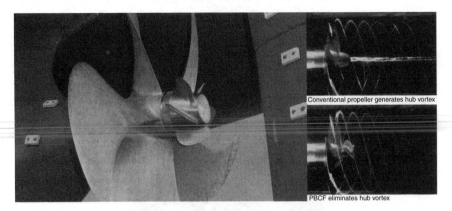

Conventional propeller generates hub vortex

PBCF eliminates hub vortex

Fig. 22.8   The PBCF concept.

- Saving of 4–5% in fuel consumption or 1–2% boost in ship speed.
- 3% reduction in propeller torque ratio.
- Reduced vibration due to the hub vortex.

PBCF can be designed for installation on ships of any type or size. They can be fitted on newly built ships, as well as those already in service. To date, 330 ships have been retrofitted with this unit.

Voith cycloidal rudder

This is a new propulsion and manoeuvring system for ships. This mechanism is placed behind a conventional propeller revolving on a horizontal axis. It in fact replaces (very efficiently) a conventional aerofoil rudder. This modern concept can be fitted on single-screw or twin-screw vessels.

It is based on the vertical axis cycloidal propeller (VSP) described earlier in this chapter. As with the VSP design, the Voith cycloidal rudder (VCR) has a rotor casing with a vertical axis of rotation (see Figure 22.9). Two rudder blades lying parallel to the axis of the rotor casing project from it.

The main advantages of replacing a conventional rudder with a VCR unit are:

- Low-resistance rudder for high-speed operation.
- Improved manoeuvrability in comparison to the conventional propulsion arrangement.
- As the cycloidal rudder is the main propulsion for low speeds, controllable-pitch propellers may be replaced by fixed-pitch propellers.
- Redundancy of propulsion and steering, thus improving 'take home' capability.
- Roll stabilisation even for a stationary vessel is possible.
- High-shock resistance, low-magnetic signature, low-radiated noise levels and thus are excellent for Royal Naval ships.

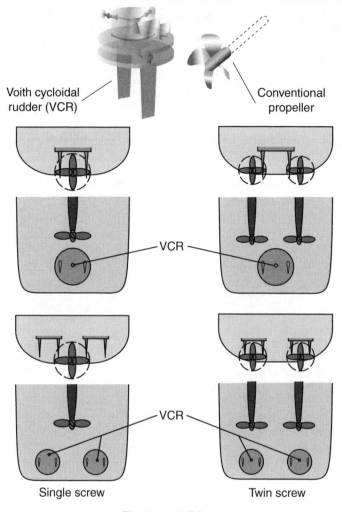

Fig. 22.9   VCRs.

VCR designs are an ideal complement to advanced propulsion systems. They could cause conventional rudders to become obsolete.

Pods

These form part of a propulsion system called COmbined Diesel-Electric and Diesel-mechanical (CODED) units. As can be seen in Figures 22.10 and 22.11, it can be a 'hybrid power' system. On the centreline is a conventionally driven controllable-pitch propeller. Port and Starboard (P&S), there are two electrically driven pods suspended and controlled from the Steering Gear Compartment.

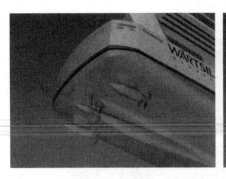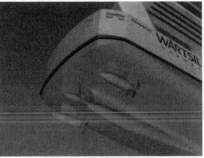

Fig. 22.10   An impression of the CODED hybrid propulsion system in action: in high-speed mode (left) and low-speed mode (right) with the centre propeller feathered.

Fig. 22.11   Two-wing pods to assist the main propeller.

These pods may be fixed in a fore and aft direction. They can also be designed to be azimuth pods, capable of operating through 360°. Note how the propellers on these P&S pods face forward. They are generally fitted on high-speed cruise ships. The QM2 (on her maiden voyage in January 2004) had four of these pods fitted. Two were fixed and two were of azimuth design.

The advantages of fitting pods are as follows:

1. It does away with the need to have aft transverse thrusters.
2. Obviates the need to have twin-screw shafting, thus saving in first costs and fuel costs.
3. Compared to a twin-screw arrangement, they reduce the resistance or drag of the ship.

4. They save volume within the ship that a twin-screw arrangement would require.
5. As a secondary power, they can assist the main propeller for forward speed.
6. At low speeds with main propeller stopped, the ship can be driven ahead by the two pods.
7. Instead of the usual 85% Maximum Continuous Rating (MCR), the two pods can operate continually at 100% rating.
8. With a forward transverse thruster, they are excellent for moving a ship bodily from and towards a jetty wall.
9. Pods fitted P&S are splendid for ship manoeuvrability in confined waters.
10. Easy to maintain and repair.
11. When fitted P&S, the pods are operating in much better wake conditions than when fitted at the centreline.

An alternative design is to have a pod fitted on the centreline aft of the main propeller. The propeller on the pod again faces forward It interacts with the main propeller to increase the overall main propeller's efficiency.

One slight disadvantage with pods is that in heavy weather conditions the blades on the pods can be bent or broken, resulting in a loss of power.

Steerpropulsion contra-rotating propellers

These designs have been fitted on offshore supply vessels and inland cruise ships. Two 120 m vessels built by De Hoop, the Netherlands were each fitted in 2001 with these propulsion units. Aft propeller can be up to 2.7 m diameter, with the forward propeller being up to 3.3 m diameter.

Conclusions

The foregone notes indicate many ideas for improving the performance of ships. However, there must be a note of caution.

When listening to claims of increases in performance, one should ascertain who is making the claim. Is it the salesman? Is it the academic or is it the industrial researcher? Each one could give a different opinion.

A lot of new designs work well on ship models and small ships. However, on larger vessels the technology breaks down or the costs for providing the technology is too prohibitive. It is similar to designing or requesting a gold-plated Mini.

There is one other point to carefully consider. Over a length of time, the percentage return against effort decreases exponentially.

Useful design and performance formulae

Preliminary estimates for dimensions

$C_D = dwt/W$ where $W = lwt + dwt$

$$C_B = \frac{\text{Volume of displacement}}{L \times B \times d}$$

$$L = \left[\frac{dwt \times (L/B)^2 \times (B/H)}{p \times C_B \times C_D} \right]^{1/3}$$

where:

$B = (L/10) + (5.0–7.5)$, for General Cargo ships
$B = (L/10) + (7.5–10.0)$, for Container ships
$L/B = 5.00 – 5.75$, for modern Supertankers
$C_B = 1.2 - (0.39 \times V/L^{0.5})$
$L = 5.32 \times dwt^{0.351}$ approximately for General Cargo ships

Estimates for steel weight

$W_d = W_b \times (W_d/W_b) \times (L_d/L_b)$

$$\text{Average sheer} = \frac{\text{Sheer Aft} + \text{Sheer Forward}}{6}$$

$$W_{ST} = 26.6 \times 10^{-3} \times L^{1.65} \frac{(B + D + H/2)(0.5C_B + 0.4)}{0.8}$$

Estimates for wood and outfit weight

$$\alpha = \text{W\&O weight for basic ship} \times \frac{100}{L_B \times B_B}$$

$$\text{W\&O weight} = \frac{1}{2}(\text{W\&O})_B + \frac{1}{2}(\text{W\&O})_B \times \frac{L_D \times B_D}{L_B \times B_B} \qquad \text{Cargo ships}$$

$$W\&O \text{ weight} = \frac{2}{3}(W\&O)_B + \frac{1}{3}(W\&O)_B \times \frac{L_D \times B_D}{L_B \times B_B} \qquad \text{Oil Tankers}$$

Estimates for machinery weight

$$A_C = \frac{W^{2/3} \times V^3}{P} \qquad \text{if } V < 20 \text{ kt}$$

$$A_C = \frac{W^{2/3} \times V^4}{P} \qquad \text{if } V \text{ equals or} > 20 \text{ kt}$$

$$M_W = 0.075\,P_B + 300 \qquad M_W = 0.045\,P_S + 500 \qquad M_W = 10.2\,P_S^{0.5}$$

Estimates for capacities

Grain Capacity = Mld Capacity × 98% approximately

Bale Capacity = Grain Capacity × 90% approximately

Insulated Capacity = Mld Capacity × 75% approximately

$$G_D = G_B \times \frac{L_D \times B_D \times 'D_D' \times C_B@SLWL_D}{L_B \times B_B \times 'D_B' \times C_B@SLWL_B}$$

$$'D' = \text{Depth Mld} + \frac{\text{Camber}}{2} + \frac{\text{Sheer Aft} + \text{Sheer Forward}}{6}$$

$$- \text{Tank Top height} - \text{Tank Top ceiling}$$

$$V_t = L_t \times B \times D_t \times C_B \times 1.16 \qquad \text{for Tankers}$$

$$V_t = L_h \times B \times D_h \times C_B \times 1.19 \qquad \text{for Bulk Carriers}$$

$$C_B@85\% \text{ Depth Mld} = C_B@SLWL \times 101.5\% \text{ approximately}$$

Approximate hydrostatics

$$\text{Any } C_B = C_B@SLWL \times \left(\frac{\text{Any waterline}}{\text{SLWL}}\right)^x$$

$$x = 4.5 \times e^{-5 \times C_B@SLWL} \qquad \text{where } e = 2.718$$

$$K = \frac{1 - C_B}{3}$$

$$C_W = (2/3 \times C_B) + 1/3 \qquad \text{at SLWL}$$

$$W = L \times B \times H \times C_B \times p$$

$$KB = 0.535 \times H \qquad 2/3 \times H \qquad H/2 \qquad \text{for various ship types}$$

$$KB = \frac{H}{1 + C_B/C_W} \qquad GM_L = BM_L \text{ approximately}$$

$$BM_T = I_T/V \qquad\qquad BM_L = I_L/V$$

$$BM_T = \frac{\eta_T \times B^2}{H \times C_B} \qquad\qquad BM_L = \frac{\eta_L \times L^2}{H \times C_B}$$

$$\eta_T = 0.084 \times C_W^2 \qquad \text{or} \qquad \eta_T = 1/12 \times C_W^2 \text{ approximately}$$

$$\eta_L = 0.075 \times C_W^2 \qquad \text{or} \qquad \eta_T = 3/40 \times C_W^2 \text{ approximately}$$

$$KM_T = KG + GM_T \qquad\qquad KM_T = KB + BM_T$$

$$WPA = L \times B \times C_W \qquad\qquad TPC_{SW} = WPA/97.56$$

$$MCTC_{SW} = 7.8 \times (TPC_{SW}^2)/B \qquad \text{for Oil Tankers}$$

$$MCTC_{SW} = 7.31 \times (TPC_{SW}^2)/B \qquad \text{for General Cargo ships}$$

$$WPA = K \times H^n \qquad H_2/H_1 = (W_2/W_1)^{C_B/C_W}$$

At each draft $BM_L/BM_T = 0.893 \times (L/B)^2$

Ship resistance

$R_t = R_f + R_r \qquad \text{where } R_f = f \times A \times V^n$

$$f_s = \frac{0.441}{L_S^{0.0088}} \qquad f_m = \frac{0.6234}{L_m^{0.1176}}$$

$$WSA_{Taylor} = 2.56 \times (W \times L)^{0.5} \qquad \frac{V_S}{L_S^{0.5}} = \frac{V_m}{L_m^{0.5}}$$

$$Fn = \frac{V}{(g \times L)^{0.5}}$$

$$\frac{R_r(\text{ship})}{R_r(\text{model})} = \left(\frac{L_S}{L_m}\right)^3 \qquad R_r \propto L^3$$

$R_r \propto$ Volume of displacement   Areas $\propto L^2$

Velocity $\propto$ (Volume of displacement)$^{1/6}$

$Rf_m \propto L_m^{2.7949}$ for ship models

$Rf_s \propto L_s^{2.9037}$ for full size ships

Types of ship speed

$V_T = P \times N \times 60 / 1852 = P \times N / 30.866$

Apparent slip ratio $= \dfrac{V_T - V_S}{V_T}$ Real slip ratio $= \dfrac{V_T - V_a}{V_T}$

$W_t = \dfrac{V_S - V_a}{V_S}$ $W_t = \dfrac{C_B}{2} - 0.05$ approximately

Pitch ratio $= \dfrac{\text{Propeller pitch}}{\text{Propeller diameter}}$

Types of power

$P_T = \text{Thrust} \times V_a$ $P_D = 2 \times \pi \times N \times T$ $P_{NE} = R_T \times V_S$

$P_E = P_{NE} + (\text{weather and appendage allowances})$

$P_E/P_T = \text{Hull efficiency}$ $P_T/P_D = \text{Propeller efficiency}$

P_D/P_B or $P_D/P_S = \text{Propeller shaft efficiency}$

P_B/P_I or $P_S/P_I = \text{Engine's mechanical efficiency}$

$P_I = X/Y$

where:

$X = P_{NE} + (\text{weather and appendage allowances})$
$Y = (\text{Hull efficiency}) (\text{Propeller efficiency}) (\text{Propeller shaft efficiency}) (\text{Engine efficiency})$

Power coefficients

$QPC = P_E/P_D$ $PC = P_E/P_B$ or P_E/P_S

$QPC = 0.85 - \dfrac{N \times L^{0.5}}{10\,000}$ approximately

$A_C = \dfrac{W^{2/3} \times V^3}{P}$ if $V < 20$ kt

$A_C = \dfrac{W^{2/3} \times V^4}{P}$ if V equals or >20 kt

$A_C = 26(L^{0.5} + 150/V)$ approximately

Propeller and rudder design

Thickness fraction $= t/D$ $a = $ Pitch/Diameter

$$BAR = \frac{\text{Total blade area}}{\pi \times d^2/4} \qquad B_p = \frac{0.0367 \times N \times P^{0.5}}{V_a^{2.5}}$$

$$\delta = 3.28 \times N \times d / V_a$$

$$\text{Thrust in N/cm}^2 = \frac{\text{Thrust in Newtons}}{\text{Total blade area}}$$

$$A_R = K \times LBP \times d \qquad F = \beta \times A_R \times V^2$$

$$F_t = Fn \cos \alpha = F \sin \alpha \, \cos \alpha$$

$$F_t = \beta \times A_R \times V^2 \times \sin \alpha \cos \alpha$$

Bollard pulls

Total required bollard pull $= (60 \times W/100\,000) + 40$

Total required bollard pull $= (0.7 \times LBP) - 35$

$$\text{ASD: bollard pull} = 0.016 \times P_B \qquad \text{VS: bollard pull} = 0.012 \times P_B$$

Speed Trials

$$\text{True speed} = \frac{V_1 + 3V_2 + 3V_2 + V_4}{8}$$

$$\text{True speed} = \frac{V_1 + 5V_2 + 10V_3 + 10V_4 + 5V_5 + V_6}{32}$$

$$\text{True speed} = N \times \frac{60}{Nm}$$

Fuel consumption trials

$$\text{Fuel cons/day} = \frac{W^{2/3} \times V^3}{F_C}$$

where:

$F_C = 110\,000$, for Steam Turbine machinery
$F_C = 120\,000$, for Diesel machinery
$F_C = 0.20\,\text{kg/kW h}$ ($0.0048 \times P_S$ tonnes/day), for Steam Turbines machinery
$F_C = 0.18\,\text{kg/kW h}$ ($0.00432 \times P_B$ tonnes/day), for Diesel machinery

Crash-stop manoeuvres

$$S = 0.38 \left[\frac{dwt^2 - dwt}{100\,000} \right] + 1.60$$

$$T = 2.67 \left[\frac{dwt}{100\,000} \right]^2 - 0.67 \left[\frac{dwt}{100\,000} \right] + 10.00$$

$$S_L = 2 \left[\frac{dwt}{100\,000} \right] + 10.50$$

Ship squat

$$\delta_{max} = \frac{C_B \times S^{0.81} \times V^{2.08}}{20} \qquad y_2 = y_0 - \delta_{max} \qquad y_0 = H - T$$

$$\delta_{max} = \frac{C_B \times V^2}{100} \text{ Open water} \qquad \delta_{max} = \frac{C_B \times V^2}{50} \text{ Confined channel}$$

$$K = (6 \times S) + 0.4 \qquad S = \frac{b \times T}{B \times H}$$

$$K_t = 40(0.700 - C_B)^2 \qquad K_{o/e} = 1 - 40(0.700 - C_B)^2$$
$$K_{mbs} = 1 - 20(0.700 - C_B)^2$$

Dynamic trim $= K_t \times \delta_{max} \qquad \delta_{o/e} = K_{o/e} \times \delta_{max} \qquad mbs = K_{mbs} \times \delta_{max}$
% loss in speed $= 60 - (25 \times H/T)$
% loss in revolutions $= 18 - (10/3 \times H/T)$
% loss in speed $= (300 \times S) - 16.5\,\%$
% loss in revolutions $= (24 \times S) + 11.6$

Reduced speed and loss of revolutions

$$F_B = 7.04/C_B^{0.85} \qquad F_D = 4.44/C_B^{1.3}$$

% loss in speed $= 60 - (25 \times H/T)$
% loss in revolustions $= 18 - (10/3 \times H/T)$
% loss in speed $= (300 \times S) - 16.5$
% loss in revolutions $= (24 \times S) + 11.6$

Interaction

$$S = \frac{(b_1 \times T_1) + (b_2 \times T_2)}{B \times H}$$

% increase in squat $= 150 - (10 \times V)$

Ship vibration

$$N = \frac{1}{T} \qquad N_{\text{Schlick}} = \emptyset\left[\frac{I_{\text{NA}}}{W \times L^3}\right]^{0.5} \qquad N_{\text{Todd}} = \beta\left[\frac{B \times D}{W \times L^3}\right]^{0.5}$$

$$W_2 = W\left[\frac{B}{3 \times d} + 1.2\right]$$

$$N_{\text{Todd and Marwood}} = \left[2.29 \times 10^6 \times \frac{I_{\text{NA}}}{(W_2 \times L^3)^{0.5}}\right] + 28$$

$$N_{\text{Burrill}} = 4.34 \times 10^6 \times \left[\frac{1+B}{2 \times d}(1+r_s)\right]^{-0.5} \times \left[\frac{I_{\text{NA}}}{W \times L^3}\right]^{0.5}$$

$$\emptyset_{\text{Schlick}} = 3.15 \times 10^6 \times C_B^{0.5} \text{ approximately}$$

$$\beta_{\text{Todd}} = 124\,000 \times C_B^{0.6} \text{ approximately}$$

$$\text{2nd harmonics} = \frac{\text{1st harmonics}}{\text{Number of blades on propeller}}$$

Revision one-liners for student's examination preparation

The following one-line questions will act as an aid to examination preparation. They are similar in effect to using mental arithmetic when preparing for a mathematical examination.

Elements of these questions may well appear in the written papers, coursework or in the oral examinations. Should you have temporally forgotten, a quick recap of the appropriate chapter notes will remind you of the answer … Good luck!!

List four items of information given by the owner to the builder for a new ship.
What is the air-draft on a ship?
List the items in the lightweight of a ship.
List six items in the deadweight of a ship.
Give the formulae for C_B and C_D.
What is a balance of weights table for a vessel?
Discuss briefly the development of Alexander's formulae for a ship's C_B.

For steel weight prediction, discuss the Cubic Number method.
For steel weight prediction, discuss the Method of differences.
What will be the percentage for length provided 30% for depth, 55% for breadth.
List the items within the Wood and Outfit weight for a ship.
In future years, why will the Wood and Outfit weight decrease for new ships?
List some non-ferrous metals used for ship structures.
Why are there two formulae for the Admiralty Coefficient (A_C)?
Why are plastics used on ships?
Give the formula for M_W for approximating Diesel machinery weight.
Give the formula for M_W for approximating Steam Turbine machinery weight.

What is the link between Moulded Capacity and Grain Capacity?
What is the link between Grain Capacity and Bale Capacity?
What is the relationship between Moulded Capacity and Insulated Volume on a 'Reefer'.

How is the capacity depth D_C evaluated?

At SLWL the C_B is 0.800. What is the approximate C_B at 85% Depth Mld?

On Oil Tankers, what exactly is the length L_t?

On Oil Tankers, what exactly is the depth D_t?

Discuss and give the modification coefficient for hull form on Oil Tankers.

Discuss and give the modification coefficient for hull form on Bulk Carriers.

For C_B at any draft below the SLWL, what is the value of 'x'?

Give the formula for predicting the C_B at any draft below the SLWL.

Give three formulae for evaluating the value KB.

What are the formulae for the transverse and longitudinal inertia coefficients?

What is the Metacentre KM_T and what is the Metacentric height GM_T?

What are the formulae for WPA and TPC?

Give typical GM_T values for three ship types when in fully loaded condition.

List the four components of total ship resistance R_T.

What is W. Froude's formula for frictional resistance R_f?

What are the 1991 formulae for f_m and f_s?

What is the Froude's speed–length law?

What is a Froude's No.?

R_f varies as L^x for geosim ship models. What is the value of x?

How do residual resistances vary with length?

R_f varies as L^x for geosim ships. What is the value of x?

Define the speeds V_T, V_S and V_a.

What are apparent slip and real slip?

What is a wake fraction weight?

Give a range of values for apparent slip and real slip.

If C_B is 0.722, then estimate the corresponding W_t value.

Which two powers are located at the thrust-block?

Use a sketch to show the positions of six powers along a propeller shaft.

What is the naked effective power (P_{NE})?

Give typical formulae for ship's hull efficiency and engine's mechanical efficiency.

What are the formulae for thrust power and delivered power?

Why are power coefficients, as used by Naval Architects?

Give two formulae for the QPC.

Which power coefficient links effective power and power located at the thrust-block?

$V = 158\,kt$, $P_B = 495$ and $W = 14\,400$ tonnes. Calculate the A_C.

$W = 16\,125$ tonnes, $V = 23\,kt$ and $P_S = 13\,610\,kW$. Calculate the A_C.

List the information shown on a B_p propeller chart.

What is a propeller's BAR?

What is the formula for B_p in terms of N, P_D and V_a?

Of what significance is the optimum pith-ratio line on a B_p chart?

Suggest a range of efficiencies for a ship's four-bladed propeller.
What is propeller cavitation?
How is the rudder value A_R calculated?
What is the rudder value K for an Oil Tanker?
Rudders perform two functions. What are they?
Sketch a Mariner-type rudder.
Whereabouts is a rudder horn?
What is the formula for the rudder force F_t in newtons?

Ship Trials can be split into four groups. Name each group.
What is a typical difference in value between trial speed and service speed?
How many metres form a nautical mile?
What is the true speed for a ship after four runs spaced equal time apart?
Sketch a graph of RPNM against time of day in middle of run.
Sketch two graphs to illustrate slack water conditions for a tide.
List four items of information measured on Consumption Trials.
What is a typical fuel in kg/kW-h for Steam Turbine machinery?
What is a typical fuel in kg/kW-h for Diesel machinery?
W = 232.000 tonnes, V = 15.1 kt, P_B = 25 125 kW. Estimate fuel consumption/day in tonnes.

Give the fuel consumption coefficient F_C for Steam Turbine machinery.
List three procedures carried out on Manoeuvring Trials.
What is the 'overshoot' in a Zig-zag Trial?
In terms of LBP, what can be the value of the turning circle diameter (TCD)?
On Crash-stop manoeuvres, why do Diesel engines give the better results?
List three precautions to be considered prior to taking a ship on trials.
What answer does $(N \times 60)/Nm$ give?
How may the time on the measured mile be measured?
Give one reason why a new vessel may not obtain her predicted trial speed.

What exactly is ship squat?
Why has ship squat become so important in the last 40 years?
List four signs that a vessel has entered shallow waters.
What is a blockage factor?
Give the overall formula for predicting maximum ship squat in metres.
Give two shortcut formulae for predicting maximum ship squat in metres.
What are the advantages of being able to predict maximum ship squat?
In the study of ship squat, to what does H/T refer?
What is the Width of Influence (F_B) and what is the depth of influence (F_D)?
What are the modern formulae for F_B and F_D?
Whereabouts will the maximum squat in shallow waters occur, if a vessel when static has trim by the stern?
What is the best way of reducing ship squat in shallow waters?

What exactly are 'Interaction' effects?
Define with the aid of a sketch, a ship's domain.
What are pressure bulbs around a moving vessel?

List the possible Interaction problems as two ships cross in a narrow river. Show how Interaction can cause a small vessel to be bowled over by a larger vessel.

List three methods of decreasing the effects of Interaction in a narrow river.

In a vibrating mass, what are nodes, anti-nodes and modes?

What is resonance or synchronisation?

Give the other name for 'entrained water'.

List three causes of vibration on ships.

List three methods for reducing vibration on existing ships.

Give the 2NV mode frequency in cycles/min formula suggested by F. Todd.

Give the 2NV mode frequency in cycles/min formula suggested by Todd and Marwood.

With regard to ship vibration, what does '3 mm out to out' mean?

List the differences between a Kort nozzle and a Kort rudder.

List the advantages of fitting a bulbous bow to a ship.

Why are double-skin hulls fitted on Oil Tankers?

What are activated stabilising tanks?

At what ship speed is a transverse thruster most efficient?

What are Grouthues–Spork spoilers?

Suggest the best method for reducing excessive rolling of a ship at sea.

Why may a Tee-duct be fitted in a Fore Peak Tank?

In ship propulsion mechanisms, what are azimuthing pods?

Suggest three areas for future research into ship-handling or propulsion mechanisms.

How to pass examinations in Maritime Studies

To pass exams you have to be like a successful football or hockey team. You will need:

Ability, Tenacity, Consistency, Good preparation and Luck!!

The following tips should help you to obtain extra marks that could turn that 36% into a 42% + pass or an 81% into an Honours 85% + award – Good luck.

Before your examination

1. Select 'bankers' for each subject. Certain topics come up very often. You will have certain topics fully understood. Bank on these appearing on the exam paper.
2. Do not swat 100% of your course notes. Omit about 10% and concentrate on the 90%. In that 10% will be some topics you will never be able to understand fully.
3. Work through passed exam papers in order to gauge the standard and the time factor to complete the required solution. Complete and hand in every set Coursework assignment.
4. Write all formulae discussed in each subject on pages at the rear of your notes.
5. In your notes circle each formula in a red outline or use a highlight pen. In this way they will stand out from the rest of your notes. Remember formulae are like spanners. Some you will use more than others but all can be used to solve a problem.
6. Underline in red important key phrases or words. Examiners will be looking for these in your answers. Oblige them and obtain the marks.
7. Revise each subject in carefully planned sequence so as not to be rusty on a set of notes that you have not read for some time whilst you have been sitting other exams.
8. Be aggressive in your mental approach to do your best. If you have prepared well there will be less nervous approach and like the football team you will gain your goal.

In your examination

1. Use big sketches. Small sketches tend to irritate Examiners.
2. Use coloured pencils. Drawings look better with a bit of colour.
3. Use a 150 mm rule to make better sketches and a more professional drawing.
4. Have big writing to make it easier to read. Make it neat. Use a pen rather than a biro. Reading a piece of work written in biro is harder to read especially if the quality of the biro is not very good.
5. Use plenty of paragraphs. It makes it easier to read.
6. Write down any data you wish to remember. To write it makes it easier and longer to retain in your memory.
7. Be careful in your answers that you do not suggest things or situations that would endanger the ship or the onboard personnel.
8. Reread your answers near the end of the exam. Omitting the word *not* does make such a difference.
9. Reread your question as you finish each answer. For example, do not miss part (c) of an answer and throw away marks you could have obtained.
10. Treat the exam as an advertisement of your ability rather than an obstacle to be overcome. If you think you will fail, then you probably will fail.

References

Baker, G.S. (1951) *Ship Design, Resistance and Screw Propulsion*, Birchall.

Barrass, C.B. (1977) *A Unified Approach to Ship Squat*, Institute of Nautical Studies.

Barrass, C.B. (1978) *Calculating Squat – A Practical Approach*, Safety at Sea Journal.

Barrass, C.B. (2003a) *Ship Squat – 32 Years of Research*, Research paper.

Barrass, C.B. (2003b) *Ship Squat – A Guide for Masters*, Research paper.

Barrass, C.B. (2003c) *Widths and Depths of Influence*, Research paper.

Barrass, C.B. (2003d) *Ship Stability for Masters and Mates*, Elsevier Ltd.

Barrass, C.B. (2003e) *Ship Stability – Notes and Examples*, Elsevier Ltd.

Carlton, J.S. (1994) *Marine Propellers and Propulsion*, Elsevier Ltd.

Eyres, D.J. (2001) *Ship Construction*, Elsevier Ltd.

Fairplay Contributors (1980) *Standard Ships (General Cargo Designs)*, Fairplay.

Hensen, H. (1997) *Tug Use in Port – A Practical Guide*, Institute of Nautical Studies.

Jurgens, B. (2002) *The Fascination of the Voith Scheider Propulsion*, Koehlers Verlagsgesellschaft mbH.

Lackenby, H. (1963) *The Effect of Shallow Water on Ship Speed*, Shipbuilder and Marine Engineering Builder.

McGeorge, H.D. (1998) *Marine Auxiliary Machinery*, Elsevier Ltd.

Moltrecht, T. (2002) *Development of the Cycloidal Rudder (VCR)*, SNAME.

Munro-Smith, R. (1975) *Elements of Ship Design*, Marine Media Management Ltd.

Nautical Studies Institute (1993–2004) *Seaways* – Monthly Journals, Institute of Nautical Studies.

Patience, G. (1991) *Developments in Marine Propellers*, Institute of Mechanical Engineers.

Rawson, K.J. and Tupper, E.C. (2001) *Basic Ship Theory*, Elsevier Ltd.

RINA (1993–2003) *Significant Ships* – Annual Publications, RINA.

RINA (1993–2004) *The Naval Architect* – Monthly Journals, RINA.

Schneekluth, H. and Bertram, V. (1998) *Ship Design for Efficiency and Economy*, Elsevier Ltd.

Stokoe, E.A. (1991) *Naval Architecture for Marine Engineeers*, Thomas Reed Ltd.

Taylor, D. (1996) *Introduction to Marine Engineering*, Elsevier Ltd.

Todd, F. (1962) *Ship Hull Vibration*, Arnold.

Tupper, E.C. (1996) *Introduction to Naval Architecture*, Elsevier Ltd.

Watson, D.G.M. (1998) *Practical Ship Design*, Elsevier Ltd.

Yamaguchi et al. (1968) *Full Scale Tests on Sinkage of a Supertanker*, Nautical Society of Japan.

Answers to questions

Chapter 1

1. L = 203 m, B = 32.58 m, displacement = 67 735 tonnes, lightweight = 12 735 tonnes.
2. Review chapter notes.
3. L = 147.3 m, B = 21.45 m, C_B = 0.734, W = 19 615 tonnes.
4. L = 265.6 m, B = 44.26 m, T = 14.167 m.
5. (a) L = 101.90 m, (b) L = 127.39 m, (c) L = 145.86 m.
6. C_B = 0.575 and also 0.575 (via global formula).

Chapter 2

Section 1
1. to 3. Review chapter notes.
4. 2900 tonnes.
5. 4350 tonnes.
6. Review chapter notes.

Section 2
1. to 4. Review chapter notes.
5. (a) 748 or 760 tonnes, with average = 754 tonnes, (b) Review chapter notes.

Section 3
1. Review chapter notes.
2. 465.
3. Diesel = 1275 tonnes, Steam Turbine machinery = 1085 tonnes.
4. 616 or 621 tonnes.
5. (a) 240 tonnes, (b) 80 tonnes.

Chapter 3
1. and 2. Review chapter notes.
3. C_B = 0.703.

4. Review chapter notes.
5. Grain $= 20\,273\,m^3$, Bale $= 18\,246\,m^3$.

Chapter 4
1. K $= 0.142, 0.125, 0.100, 0.075, 0.058, 0.050$.
2. $0.839, 0.843$.
3. (a) $0.782, 0.767, 0.748, 0.721$.
   (b) W $= 50\,195$ tonnes, lightweight $= 17\,356$ tonnes, deadweight $= 33\,559$ tonnes.
4. Review chapter notes.
5. W $= 13\,251$ tonnes, $C_W = 0.813$, $C_B/C_W = 0.886$, KB $= 3.98\,m$, WPA $= 1946\,m^2$, TPC $= 19.95$ tonnes, MCTC $= 161.1\,tm/cm$.
6. $0.813, 0.0551, 3.31, 162.5\,m$.

Chapter 5
1. Model's 'f' $= 0.5002$, ship's 'f' $= 0.4225$.
2. $3389\,m^2$.
3. $13.89\,kt$.
4. (a) Fn $= 0.248$, (b) review chapter notes.
5. Review chapter notes.
6. $33.45\,N$.
7. $81.63\,kN$.
8. $P_{NE} = 1364\,kW$, V $= 13.28\,kt$.

Chapter 6
1. Review chapter notes.
2. $16.5\,kt$.
3. (a) Review chapter notes, (b) $W_t = C_B/2 - 0.05$.
4. $V_s = 10.01\,kt$, apparent slip $= -11.1\%$, real slip $= +16.66\%$.

Chapter 7
1. to 4.   Review chapter notes.
5. (a) $P_E = 4118$, $P_T = 4149$, $P_D = 6035$, $P_B = 6212$, $P_I = 7001$ (all in kW).
   (b) $177\,kW$.

Chapter 8
1. PC $= 0.672$, QPC $= 0.727$.
2. $14.52\,kt$.
3. $63\,467$.
4. $21\,kt$.
5. 569 (very efficient design).
6. $14\,218\,kW$.

Chapter 9
1. to 4.   Review chapter notes.
5. B_p = 15.29, propeller efficiency = 68.7%, δ = 160, a = 0.844, pitch = 5.04 m, diameter = 5.97 m.
6. A_R = 30.6 m², L = 4.25 m, D = 7.20 m.
7. 500 kN.

Chapter 13
1. and 2.   Review chapter notes.
3. (1) 17 kt, (11) 0.08, 0.29, 0.71, 0.94, 1.11, 1.38 (all going North).
4. 12.17, 11.98, 14.24, 14.01, 14.07, 13.85 (all in kt).

Chapter 14
1. Review chapter notes.
2. 67, 92, 109 tonnes.
3. Steam Turbine = 0.00480 × P_S, Diesel = 0.00432 × P_B.
4. 14 000, 20 000, 26 250 kW.

Chapter 15
1. 2.10 nm, 14 lengths, 17 min.
2. Review chapter notes.
3. Review chapter notes.
4. 0.57 nm (¼ × S), 0.76 nm (⅓ × S), 5.70 nm (2.5 × deep water value).

Chapter 17
1. and 2.   Review chapter notes.
3. 1.41 m at the bow.
4. 0.71 m at the bow.
5. (a) 0.89 m at the stern, (b) 1.78 m at the stern.
6., 7. and 8.   Review chapter notes.
9. Bow = 0.40 m, stern = 0.84 m, mbs = 0.62 m, trim = 0.44 m by the stern.

Chapter 18
1. Review chapter notes.
2. F_B = 351 m, F_D = 54 m.
3. 29.7% or 4.46 kt.
4. 17% or 19 rpm.

Chapter 19
1. to 3.   Review chapter notes.
4. 0.225.
5. Review chapter notes.

Chapter 20

1. Schlick's constant $= 2.719 \times 10^6$, Todd's constant $= 103\,924$.
2. to 4.   Review chapter notes.
5. 2NV $= 73.04$ cycles/min, 2NH $= 110$ cycles/min.
6. 70.85 cycles/min.
7. Review chapter notes.

Index

Page numbers in *italics* refer to figures and tables.